Human Identity and Bioethics

When philosophers address personal identity, they usually explore numerical identity: What are the criteria for a person's continuing existence? When nonphilosophers address personal identity, they often have in mind narrative identity: Which characteristics of a particular person are especially salient to her self-conception? This book develops accounts of both senses of identity, arguing that both are normatively important, and is unique in its exploration of a wide range of issues in bioethics through the lens of identity. Defending a biological view of our numerical identity and a framework for understanding narrative identity, David DeGrazia investigates various issues for which considerations of identity prove critical: the definition of death; the authority of advance directives in cases of severe dementia; the use of enhancement technologies; prenatal genetic interventions; and certain types of reproductive choices. *Human Identity and Bioethics* demonstrates the power of personal identity theory to illuminate issues in bioethics as they bring philosophical theory to life.

David DeGrazia is Professor of Philosophy at George Washington University. He is the author of *Taking Animals Seriously: Mental Life and Moral Status* and *Animal Rights: A Very Short Introduction* and coeditor, with Thomas Mappes, of *Biomedical Ethics* in its fourth, fifth, and sixth editions.

Human Identity and Bioethics

DAVID DeGRAZIA

George Washington University

CAMBRIDGE
UNIVERSITY PRESS

CAMBRIDGE UNIVERSITY PRESS
Cambridge, New York, Melbourne, Madrid, Cape Town, Singapore, São Paulo

Cambridge University Press
32 Avenue of the Americas, New York, NY 10013-2473, USA

www.cambridge.org
Information on this title: www.cambridge.org/9780521825610

First published 2005

A catalog record for this publication is available from the British Library.

Library of Congress Cataloging in Publication data
DeGrazia, David.
Human identity and bioethics / David DeGrazia.
p.   cm.
Includes bibliographical references and index.
ISBN 0-521-82561-X – ISBN 0-521-53268-X (pbk.)
1. Identity (Philosophical concept)   2. Identification.   3. Human beings.
4. Bioethics.   I. Title.
BD236.D44   2005
179′.7–dc22   2004021782

ISBN 978-0-521-82561-0 hardback
ISBN 978-0-521-53268-6 paperback

Transferred to digital printing 2008

To the memory of Terry Moore, a great editor

Contents

Acknowledgments

I have been working on this book for quite a while. During this time, I have spent countless enjoyable hours reading, brainstorming, and writing. But perhaps most enjoyable of all has been the time spent exchanging ideas with academic friends.

My focused research on personal identity theory began in the summer of 1997. That summer and the following fall, while I was on sabbatical, I was a Visiting Scholar at the Institute for Philosophy and Public Policy at the University of Maryland. I am grateful to the Institute's scholars, especially David Wasserman and Robert Wachbroit, for their hospitality. At some point during my visit, I came to reject the dominant psychological approach to personal identity in favor of some type of biological approach. At around the same time, in reading Marya Schechtman's work, I recognized the importance of carefully distinguishing numerical identity, on which most analytical philosophers had focused, and narrative identity. Before long, I had come across Eric Olson's work and began to benefit from his careful defense and elaboration of the biological view of numerical identity. Subsequent communications with these two scholars were very helpful to me.

In fall 1997, I began to draft articles addressing some of the topics taken up in this book. The articles have come gradually over the years as I have tried out various ideas (and sometimes devoted myself entirely to other projects). Feedback from journals has been invaluable. Also invaluable has been feedback following talks, both formal and informal, that I have given to various audiences: one at the Institute for Philosophy and Public Policy; two at the Center for Human Values, Princeton University; three at the Kennedy Institute of Ethics, Georgetown University; three at annual meetings of the American Society for Bioethics and Humanities; and four for my colleagues in the Department of Philosophy, George Washington University. I would like to thank everyone who listened and shared ideas with me. Special thanks to Peter Singer for setting up an exchange with Jeff McMahan in my second visit to Princeton. Since then, I have learned a lot from Jeff and his writings.

Chapters began to take shape, slowly, a few years ago. Special thanks to Ray Martin, Maggie Little, and Andy Altman for encouraging responses to my initial plan for the book. As chapters were drafted and redrafted, I received written feedback from quite a few people. David Wasserman heroically read Chapters 2–7 as each emerged in embryonic form. Maggie Little responded with great insight to Chapters 5 and 7, as did Marya Schechtman to Chapters 3 and 5. Ray Martin and Jeff McMahan helpfully commented on Chapter 2. I am much obliged to Jeff Brand-Ballard for his reactions to Chapters 4–6. Many thanks also to Patricia Greenspan (Chapter 3); Madison Powers, Tom Beauchamp, and especially Robert Veatch (Chapter 4); Jeff Blustein, Ken Schaffner, LeRoy Walters, Rebecca Dresser, and Robert Wachbroit (Chapter 5); Eric Juengst, Carl Elliott, Eric Saidel, Buddy Karelis, and Ilya Farber (Chapter 6); and Dan Brock (Chapter 7). My colleagues in the Department of Philosophy also provided helpful oral feedback on the last two chapters. Finally, two anonymous reviewers for Cambridge University Press generously commented on the entire manuscript; one had responded earlier in great detail to sample chapters when I was seeking a contract.

Terry Moore, philosophy editor at Cambridge University Press, solicited this review of sample chapters and two reviews of the project prospectus. After Cambridge gave me a contract, Terry responded encouragingly to my occasional progress reports until illness forced him to delegate some duties to Stephanie Achard. Stephanie served admirably in Terry's stead in her remaining time with the Press; recently, and shortly after Terry's death (at much too young an age), Beatrice Rehl ably assumed the post of philosophy editor. During the transition, Glenna Gordon has provided helpful continuity as editorial assistant. My heartfelt thanks to everyone at Cambridge and especially to Terry, whom I will remember as someone who long ago gave a young guy a chance (leading to the publication of *Taking Animals Seriously*).

Over the past year, a sabbatical leave from George Washington University has enabled me to work steadily on the project. A fellowship from the National Endowment for the Humanities permitted me to extend the leave to two semesters without a drastic cut in pay; the Columbia

College of Arts and Sciences and, in particular, Dean Bill Frawley took care of the details. My thanks to NEH and GWU for their support; also to Paul Churchill and Bill Griffith, successive Chairs of the Department of Philosophy, for theirs; and finally to Robert Veatch and Jeff McMahan for writing letters in support of my NEH application.

Closer to home, I thank my entire family and my friends for their love. I am especially grateful to two individuals. Kathleen Smith, my wife and partner, gave me good counsel at various stages of this project. Her reminders not to hurry – and not to allow work to take over my spirit – have been enormously helpful. Meanwhile, my daughter, Zoë, provides much of the inspiration for everything I do.

1

Introduction

You and I are persons. More specifically, we are human persons – persons who are members of the species *Homo sapiens*. But what does it mean to say that someone is a person? And what is the significance of being human?

You and I have existed for years. We will continue to exist in the future. What are the criteria for our continuing to exist over time?

In continuing to exist – that is, in living our lives – we develop stories about ourselves. These stories may go well or badly from our individual perspectives. What is the character of these self-stories? At a general level, how do we want them to go? Does our existence have any value to us if we are incapable of telling such stories to ourselves?

When do we come into being, and when do we die? What is the relationship between our origins and death, on the one hand, and the boundaries of our self-stories, on the other? What are we most fundamentally? Are we essentially self-narrating persons or are we essentially human animals – who happen to treasure the portion of our lives when we can make narratives?

As persons, we not only exist over time and develop self-narratives; we plan for the future in the hope that our stories will go a certain way. But common sense suggests that we can plan for times when we no longer have the ability to plan or make any complex decisions. Advance directives in medicine are supposed to facilitate such planning. But the person who completes an advance directive may be very different from the patient to whom it will later apply. We tend to think that the earlier person and the later patient are the same individual, who has changed greatly but not gone out of existence. Is this correct? In any case, what is the authority of earlier plans for someone who no longer remembers such planning and doesn't care about it?

Self-stories and planning are characteristic of human persons. So is the effort to change ourselves in ways we consider improvements. Such changes can be minor, moderate, or extreme. Modern technologies facilitate many efforts at self-change. Are any self-transformations so drastic that they literally put one out of existence, creating someone else? Are any of them inherently unethical? Are major self-transformations via technologies, or certain technologies, morally problematic for other reasons? In the end, are they justifiable?

Future technologies will enable doctors to modify a fetus's genome, either to prevent some disease or impairment or to enhance certain traits. But would such interventions, by changing an individual's genes, effectively eliminate that individual and create a new one? Or would it merely change a persisting individual in a way that importantly affects her later self-story? Whatever the answer, would we be justified in pursuing prenatal genetic therapy or enhancement?

Today's parents routinely face reproductive decisions in light of information provided by genetic and other medical tests. Sometimes such information recommends delaying efforts to become pregnant; getting pregnant too soon would likely result in the birth of a child with a significant handicap. Suppose that a couple nevertheless seeks and achieves pregnancy immediately, predictably producing a handicapped child. Absent special circumstances, such a decision seems wrong. But the child brought into existence with a handicap would not have existed had her parents delayed conception, and so – assuming that her life is worth living – apparently lacks any basis for complaint. If the parents' choice harms no individual, how can it *wrong* anyone? And if it doesn't wrong anyone, how can it *be* wrong?

One legal option available to pregnant women is abortion. But if we come into existence as fetuses, does that mean that fetuses have full moral status and a right to life? Will any of the ideas that emerge in our investigation help to resolve this most controversial of moral problems?

The present book addresses all of these questions. The remainder of this chapter will sketch the conception of personhood with which the book will work before outlining the chapters that follow.

PERSONHOOD[1]

The word *person* traces back at least to the Latin *persona*: a mask, especially as worn by an actor, or a character or social role.[2] The concept evolved into the Roman idea of *one who has legal rights* – notably excluding slaves – before broadening into the Stoic and Christian idea of *one who has moral value.* The modern concept defines persons as *beings with the capacity for certain complex forms of consciousness,* such as rationality or self-awareness over time. Here is John Locke's classic formulation: "a thinking intelligent being, that has reason and reflection, and can consider itself, as itself, the same thinking thing, in different times and places."[3] This concept is closely associated with, and arguably includes, the idea of someone who has moral status (or full moral status if the latter comes in degrees) and perhaps also moral responsibilities.

I will take this modern concept as our shared concept of personhood. But there are different ways to sharpen it, yielding different *conceptions* of personhood. In ordinary life, when we refer to persons we are refer-ring to particular human beings. The term refers paradigmatically – that is, without controversy – to normal human beings who have advanced beyond the infant and toddler years. Such human beings are certainly beings with the capacity for complex forms of consciousness, for they are psychologically complex, highly social, linguistically competent, and richly self-aware. They are also members of our species. But must a person be human (*Homo sapiens*)? Perhaps some nonhumans display equally so-phisticated forms of consciousness. And must all members of our species qualify as persons? What about those who, although of a species whose members characteristically feature this capacity, do not themselves pos-sess it due to genetic anomaly or injury? And what is the significance of the term *capacity*? Some such term is needed to indicate that you remain a person while sleeping; you retain the relevant abilities even when not using them. But does a human fetus, currently quite unable to manifest complex forms of consciousness, have the capacity to do so in the sense of having a nature, or genetic program, that ordinarily permits development

---

[1] Parts of this section borrow significantly from my chapter "On the Question of Person-hood Beyond *Homo Sapiens*" in Peter Singer (ed.), *In Defense of Animals*, 2nd ed. (Oxford: Blackwell, forthcoming).

[2] See *The Oxford English Dictionary*, 2nd ed., vol. XI (Oxford: Clarendon, 1989), pp. 596–7.

[3] *Essay Concerning Human Understanding*, 2nd ed. (1694), Book II, ch. 27, sect. 9.

of the relevant abilities? If such potential constitutes the relevant capacity, then fetuses are persons.

Some people think that all and only members of our species are persons, regardless of their actual current abilities.[4] Others think that persons are beings who actually possess the relevant abilities, regardless of whether they are human.[5] I believe this second conception far more adequately captures the term's current *meaning*, even if in everyday circumstances people typically use the term only to *refer to* human beings. After briefly defending this claim, I will note that it is not strictly necessary for our purposes.

The concept of personhood seems to extend beyond humanity. For we often categorize as persons certain imaginary nonhuman beings and certain nonhuman beings whose existence is debatable. Thus E.T., the extraterrestrial, Spock from *Star Trek*, and the speaking, encultured apes of *The Planet of the Apes* impress us as being persons. Furthermore, if God and angels exist, they too are persons. (Interestingly, many people who are inclined to equate personhood with humanity also assert, contradictorily, that God is a person.) This suggests that *person* does not mean *human being*. The term refers to a kind of being defined by certain psychological traits or capacities: beings with particular complex forms of consciousness. So, in principle, there could be nonhuman persons, for it is conceivable – and perhaps true – that certain nonhumans have the relevant traits.

As noted, the concept of personhood is closely associated with the idea of moral status or full moral status. Is that moral idea part of the very concept of personhood? Another possibility is that the latter is purely descriptive – but seems tantamount to an assertion of moral status because virtually everyone assumes that persons have moral status (as a matter of moral principle, not linguistic meaning). But whether or not the concept of personhood combines descriptive content with moral content, it undoubtedly has descriptive content. Moreover, because the assumption that moral status requires personhood is increasingly challenged today – for example, by those who hold that sentient animals have moral status – it will be advantageous to focus on the term's less controversial, descriptive meaning.

Can we elucidate personhood in greater detail? Although many fairly specific analyses have been offered, they never seem quite right. Consider

---

4 See, e.g., Norman Ford, *The Prenatal Person* (Oxford: Blackwell, 2002), ch. 1.
5 See, e.g., Mary Anne Warren, "On the Moral and Legal Status of Abortion," *Monist* 57 (1973): 43–61 and Michael Tooley, *Abortion and Infanticide* (Oxford: Clarendon, 1983).

Harry Frankfurt's thesis that persons are beings capable of autonomy (in his terminology, *freedom of the will*): the capacity to examine critically the motivations that move one to act in a certain way, and either identify with these motivations or reject and work to change them.[6] Thus a person, or an autonomous being, may have an incessant desire to drink, due to alcoholism, but may fight this urge and seek to extinguish it. But to require so much cognitive sophistication for personhood is to require too much. No one really doubts that normal three-year-olds and moderately retarded individuals are persons, yet they may lack the capacity for critical reflection necessary for autonomy. Another view, suggested by P. F. Strawson, is that to be a person is to have both mental and bodily characteristics.[7] But surely this is too inclusive, for many animals we would never regard as persons have both types of characteristics.

Consider another definition, which is quite close to Locke's and apparently strikes many philosophers as plausible: persons as rational, self-aware beings.[8] Here the problem is that neither rationality nor self-awareness is an all-or-nothing trait. Many creatures we would not regard as persons display some rationality, which comes in degrees. For example, a cat who wants to go outside, understands that heading to the cat door will get him there, and then intentionally heads for the cat door as a means of getting outside displays simple instrumental rationality.[9] Meanwhile, self-awareness comes in different kinds as well as degrees.[10] For example, presumably all animals capable of intentional action, such as cats, have some degree of *bodily* self-awareness, an awareness of their own bodies as distinct from the rest of the environment. Relatively social mammals also have *social* self-awareness: an awareness of how they fit into group structures, expectations that come with their position in the group, likely consequences of acting against those expectations, and so on. Vervet monkeys, for example, are socially self-aware to an impressive degree.[11] A further kind of self-awareness is *introspective* awareness,

---

[6] "Freedom of the Will and the Concept of a Person," *Journal of Philosophy* 68 (1971): 829–39. Cf. Daniel Dennett, *Brainstorms* (Hassocks, England: Harvester, 1978), ch. 14.

[7] *Individuals* (London: Methuen, 1959), p. 104.

[8] Peter Singer, *Practical Ethics*, 2nd ed. (Cambridge: Cambridge University Press, 1993), pp. 110–11.

[9] I argue that many animals can act intentionally and to some degree rationally in *Taking Animals Seriously: Mental Life and Moral Status* (Cambridge: Cambridge University Press, 1996), pp. 129–72.

[10] I develop this point and discuss the relevant empirical literature, ibid., pp. 166–83.

[11] See Dorothy Cheney and Robert Seyfarth, *How Monkeys See the World* (Chicago: University of Chicago Press, 1990).

consciousness of one's own mental states. It is uncertain whether any nonhuman animals have this capacity. In any event, due to the distinct kinds of self-awareness, each of which admits of degrees, it illuminates little to say simply that personhood requires self-awareness. Which kind? If one replies that introspective awareness, say, is necessary for being a person, it would be appropriate to counter that this trait is only one of a cluster of traits that seem about equally implicated in the concept of personhood. To identify introspective awareness alone – or even in combination with rationality – as definitive of personhood would be arbitrary.

Personhood appears to be associated with a cluster of traits without being precisely analyzable in terms of any specific subset: autonomy, rationality, self-awareness, linguistic competence, sociability, the capacity for intentional action, and moral agency. A being doesn't need all these traits, however specified, to be a person, as demonstrated by nonautonomous persons. Nor is it enough to have just one of them, as indicated by the enormous range of animals capable of intentional action. A person is someone who has *enough* of these characteristics. Moreover, because we cannot draw a precise, nonarbitrary line that specifies what qualifies as enough, the concept is fairly vague. Like many or most concepts, personhood has blurred edges. Still, *person* means something, permitting us to identify paradigm persons and, beyond the easy cases, other individuals who are sufficiently similar in relevant respects to deserve inclusion under the concept.

My suggestion, then, is that the present meaning of *person* is roughly *someone (of whatever species or kind) with the capacity for sufficiently complex forms of consciousness*. I also suggest that we understand *capacity* in the sense of current capabilities; mere potential to develop them is not enough. As will become evident in Chapter 2, most leading philosophical work on personhood and personal identity agrees with this rough conception – although, as noted, scholars frequently try to sharpen it into a more specific analysis.[12]

But suppose I am mistaken in claiming that this broadly Lockean conception best expresses our concept of personhood. Or suppose (whether or not I am correct) particular readers disagree with my claim or are uncertain about it. My being mistaken or unpersuasive about the

---

[12] My comments imply that this sharpening effort is fruitless due to the concept's vagueness, assuming one is trying to capture the shared concept of personhood rather than stipulating a conception for a particular purpose.

shared concept would not matter. That is because nothing important, philosophically or morally, will turn on my usage of the term. The entire book could be read, without loss of meaning, after one has deleted every occurrence of *person* and substituted *someone with the capacity for complex forms of consciousness*. Indeed, my definition of *person* may be regarded as stipulative – as an announcement of how I intend to use the term – rather than as a thesis about the term's objective meaning. Conveniently, the vast majority of scholars I will cite use the term in ways that are consistent with my definition.

Can my use of *person* get off the argumentative hook so easily? What about the claim that, in addition to descriptive content, the term includes moral content? Here I can remain fairly neutral. The content in question involves an assertion of moral status. Not wanting to beg significant moral questions in my use of the term *person*, I will be careful in what I claim. Note that both traditional moralists, who hold that human beings have exclusive – or at least radically superior – moral status, and animal protectionists, who hold that many nonhuman animals have significant moral status, agree on this proposition: *Personhood is sufficient for full moral status*. Whether it is also *necessary* for full moral status I leave open. (I deny that it is necessary for *substantial* moral status, but that is another matter that does not affect the present discussion.[13]) The italicized statement is all I will assume, morally, about personhood. Whether it expresses part of the meaning of *person*, or states a logically independent moral assumption does not matter for our purposes. In any case, I will hereafter use *person* in the species-neutral sense articulated previously.[14]

PLAN OF THE BOOK

The remainder of this book will address the questions introduced at the beginning of this chapter. Chapters 2 and 3 provide a framework for understanding the identity of human persons. Chapter 2 – the longest and most technical in the book – confronts the issue of our *numerical identity*, which most analytic philosophers have considered *the* issue of personal identity: What are the criteria for our continuing to exist

[13] See *Taking Animals Seriously*, ch. 3.
[14] This species-neutral sense of *person* leaves open the conceptual possibility of nonhuman persons. Elsewhere I have argued that there are currently at least five living nonhuman persons – all of whom, notably, have received extensive linguistic training: three great apes and two dolphins ("On the Question of Personhood Beyond *Homo Sapiens*"). Space constraints prevent me from including this material here.

over time? Closely related is the issue of our essence: What are we most fundamentally? Against the philosophical majority, I will argue that we are essentially human animals, not minds or persons, and that our persistence conditions are biological, not psychological.

Because I deny that we are essentially persons, I will sometimes speak of *our* identity or *human* identity, rather than *personal* identity. At issue is the identity of human persons – that is, of human beings, like us, who are persons for at least part of their existence. The term *personal identity* is potentially confusing in half-suggesting either that we are essentially persons or that the issue concerns our identity only so long as we are persons. Nevertheless, the term is so well established that it would be awkward to avoid it altogether.

Chapter 3 focuses on a different sense of identity, one largely neglected by analytic philosophers. This is *narrative identity*, which involves a person's self-conception, what she considers most important to who she is, the way she organizes the story she tells herself about herself. In addition to providing a framework for understanding narrative identity, the chapter will seek to illuminate the related concepts of self-creation, autonomy, and authenticity. It will also address the issue of what most matters, prudentially, in our continued existence. Importantly, other philosophers who have focused on narrative identity have had little to say about numerical identity, apparently believing the latter unimportant. This book distinguishes itself by developing accounts of both senses of identity and maintaining that both are normatively important.

Chapters 4 to 7 engage this two-part account of human identity with specific practical issues. The most general theme uniting the chapters is that one or both senses of identity are critical to understanding a rich array of issues in bioethics: the definition of human death; the authority of advance directives in cases of severe dementia; the use of enhancement technologies; prenatal genetic interventions; and certain types of reproductive choices. With the help of plausible moral assumptions, considerations of identity illuminate these difficult issues. No less importantly, casual appeals to identity are unhelpful. Carefully distinguishing numerical and narrative identity – and having plausible views about both – are critical to identity-related argumentation in bioethics. As we will see, much of the literature conflates the two senses of the term and/or assumes implausible theses about identity, vitiating its argumentation from the start.

Chapter 4 addresses the definition of human death. Since death ends our existence, it concerns our persistence conditions – conceptually tying the issue of human death to that of our numerical identity. Currently there is a virtual consensus among scholars that the permanent cessation of functioning of the entire brain is sufficient for a human being's death. I will argue, to the contrary, that an updated version of the traditional cardiopulmonary standard best coheres with the concept of death in the case of human beings. But, because the *policy* issue of defining death cannot rely on ontological considerations alone, narrative identity and various pragmatic considerations also weigh in, leading to a more pluralistic framework. This discussion illustrates both the relevance *and the limits* of personal identity theory in addressing issues in bioethics. Sometimes good theorizing illuminates normative issues by preventing premature closure.

Both senses of identity prove important in Chapter 5, which addresses the authority of advance directives in cases of severe dementia. That our numerical identity is a function of biological life ensures that an advance directive's author remains in existence despite having even the severest dementia. At first glance, then, it would appear that advance directives carry their usual authority in such cases. But, since our persistence as self-narrators matters greatly to us, narrative identity is also salient. The investigation requires refining the framework for understanding narrative identity: Weak and strong types of narrative identity are distinguished, as are several senses of *identification*, which may or may not characterize the relationship between the earlier author of an advance directive and the later individual to whom it presumably applies. (The need to refine our theoretical framework in discussing advance directives illustrates the reciprocal dynamic of theory development in ethics: Sometimes a theoretical framework illuminates particular practical issues; sometimes the practical issues require refinement, or even revision, of the framework.) The chapter ultimately steers a middle course between those who favor precedent autonomy *über alles* and those who argue that, in cases of severe dementia, best interests trump respect for autonomy.

Chapter 6 explores enhancement technologies in relation to identity and self-creation, the deliberate shaping of oneself or one's life direction. Focusing on cosmetic surgery, cosmetic psychopharmacology, and genetic enhancements, the discussion finds most concerns about them to provide reasons for caution rather than prohibition. Most identity-related objections prove to rest either on misunderstandings concerning our identity or on a rigid romanticism about a person's current characteristics. The upshot is a cautious openness about the use of enhancement technologies in projects of self-creation.

Turning to our prenatal identity, Chapter 7 investigates prenatal genetic interventions and certain types of reproductive decisions. It is
argued that the human animal, or organism, comes into being not at
conception but somewhere between the sixteen-cell stage and the time
at which twinning becomes impossible – further refining the account of
numerical identity defended in Chapter 2. It is next argued that, once we
come into existence, our identity is relatively robust in the face of genetic
and other changes. Most arguments supporting a claim of fragile prenatal identity prove either to conflate numerical and narrative identity or
to assume that we are essentially minds or persons. Except for very early
in pregnancy, before one of us has come into being, the robustness thesis
obviates the concern that prenatal genetic interventions may put one human individual out of existence while creating a new one. Nevertheless,
for various reasons, prenatal genetic therapy enjoys somewhat stronger
moral support than prenatal genetic enhancement.

Turning to reproductive decision making, the chapter next tackles the
nonidentity problem. If a couple's choice to conceive at a particular time
predictably brings into the world a child with a handicap that could easily
have been avoided by delaying efforts to conceive, their behavior seems
highly objectionable – even if the resulting child has a worthwhile life.
The problem is to make sense of this moral judgment in light of the fact
that, had the couple delayed pregnancy, the child they would have created
is numerically distinct from the child they did, in fact, create. Since the
actual child would not have existed had the parents delayed pregnancy, he
is apparently not a victim. How, then, to explain the seeming wrongness
of the parents' behavior? I attempt to address this problem in a way that
generates neither paradox nor implausible ethical implications – and
with minimal dependence on any specific view of human identity.

The chapter's final section addresses abortion. After rebutting several
strategies for resolving this issue, including an ingenious argument from
personal identity, I reconstruct the strongest antiabortion argument –
the Future-Like-Ours Argument (which tacitly assumes the biological
view) – and appeal to the earlier-defended view of our origins to determine when in gestation this argument first applies. I then raise doubts
about a highly regarded strategy for undercutting the Future-Like-Ours
Argument before contending that an appeal to the fetus's time-relative
interests successfully defeats the argument, greatly advancing the case for
a liberal position.

2

Human Persons

Numerical Identity and Essence

Penelope was pretty sure she wanted to go through with the operation. She understood that by now, in 2061, more than 100 people had undergone body transplants. As the surgeons explained to her, this is how it would work. A healthy thirty-year-old had agreed to be a body donor if she entered a permanent vegetative state (PVS).[1] She had just been in a car accident that put her in PVS, and so, in accordance with her directive, her cerebrum was removed and discarded, making room for Penelope's. "But I thought someone in PVS wasn't legally dead," Penelope worried aloud. "You're forgetting that in our state people can opt out of the default legal definition of death and declare that PVS will be considered death in their case," a surgeon reminded her. "In her condition, the brainstem is still functional. After removing your cerebrum from your cancer-ridden body – which couldn't possibly survive for more than a few more months – we'll attach the cerebrum to her brainstem and nervous system. We're confident that within a few days of the operation you'll wake up to find yourself with a healthy new body!"

Penelope brimmed with optimism. Assuming the operation worked, someone would wake up with Penelope's functioning cerebrum and would apparently remember Penelope's life – including the decision to undergo the operation. "That person," she mused "would be me." But then a twinge of doubt hit her: "They're going to keep my original body alive for a month or so to study the effects of my cancer. If I'm in a new body, then who is that person, or being, with my old body? Is it possible

---

[1] The term "*persistent* vegetative state" is common. I prefer *permanent* because it clearly conveys irretrievable loss of the capacity for consciousness.

11

that, whatever our state laws may say, the woman who went into PVS would remain alive and acquire my cerebrum – and, with it, my personality and apparent memories? Then again, how much do these philosophical questions matter?"

Penelope's quandary raises several major questions about human persons.

1. *The question of personal identity (in the numerical sense): What are the criteria for a person's continuing to exist over time?* For any person considered at a particular time, what does it mean for that same individual to exist at another time in the past or future? Is personal identity a matter of psychological continuity, as found between Penelope before surgery and the person in the young body after surgery? Or is personal identity best understood, say, as continuity of a biological life – as exhibited between Penelope's preoperative body and the decerebrate body for a month following surgery?

2. *The question of our essence: What are we human persons, most fundamentally or essentially?* An answer to this question will help us to answer the previous one. If we are essentially persons, then we cannot exist as nonpersons, in which case our identity over time involves criteria closely related to personhood (e.g., psychological continuity). If we are essentially animals, then we can exist despite lacking the sort of psychological life associated with persons; in that case, bodily or biological criteria for our identity over time are appropriate.

3. *The question of what matters in survival: From a self-regarding or prudential – as opposed to other-regarding or ethical – standpoint, what is it about our continued existence that primarily matters?* Is it, for example, the sort of psychological continuity that allows one to maintain a sense of oneself as persisting over time? Or is it merely the capacity to have conscious experiences? Or continued existence *simpliciter*? The question of what matters in survival will be discussed in this chapter and in Chapter 3.

Focusing chiefly on the questions of personal identity and the essence of human persons, this chapter begins with an examination of the psychological view that has historical roots in Locke and became dominant in the late twentieth century primarily through Derek Parfit's work. Several strengths of the psychological view are noted, followed by concerns about the case method that supports this approach. After making the case for some form of essentialism – as opposed to a thoroughly antiessentialist view – the discussion will take up a distinct set of criticisms of the psychological view. Focusing on the question of our essence, these concerns have motivated the *constitution view*, recently developed in detail by Lynne Rudder Baker. While the constitution view can reply with some success to the aforementioned set of criticisms, this version of the psychological approach ultimately proves inadequate as a theory of our identity and essence. The chapter next presents several arguments favoring a biological view, four objections to this view, and replies to the objections. An alternative, representing a hybrid between the psychological and biological approaches, will next be examined and found deficient. A somewhat more promising view, which takes human persons to be essentially embodied minds, will also be examined and rejected. Finally, the chapter will conclude with some general reflections about our identity and essence, as well as a summary of the biological view's response to cases discussed in the chapter.

THE PSYCHOLOGICAL VIEW

Locke's Theory

What are the criteria for a person's continuing to exist over time? As it is usually understood by philosophers (later we will consider another formulation), this question is posed more specifically as follows: What makes a person at one time and a person at some other time one and the same person? This question is a special case of the ancient metaphysical puzzle of *the one and the many* – how things can persist, maintaining their identity, despite change. The relevant sense of identity or sameness here is *numerical* identity, not *qualitative* identity (or similarity). Thus, in this numerical sense, Pedro lived for many years, despite major qualitative changes over that time; and even if two twins were qualitatively identical, they would be numerically distinct, since numerical identity concerns the persistence of a single entity. Is the question of personal identity purely of metaphysical interest? Not according to Locke.

Locke responded to the apparent weaknesses of a traditional answer to the question of personal identity. According to this traditional answer, a person's identity over time consists in *sameness of substance*. Thus, for Descartes, each of us is "a thing that thinks" (*res cogitans*) – a soul or immaterial substance – so that the person lasts as long as the soul does. Many people today accept this view. Another substance-based view is possible for materialists (also called *naturalists* or *physicalists*), who hold that all substances are material or physical. This is the view that persons are physical substances, perhaps human bodies or brains, and that identity consists in the persistence of the relevant physical substance.

Locke rejects the proposal that identity consists in sameness of substance, whether immaterial or material, believing that any substance-based view will fail to account for self-knowledge regarding identity and for certain practical concerns that implicate such self-knowledge. (He rules out neither the notion of immaterial substances nor the view that "that which thinks in us" is a physical substance, just the commonly associated views about identity over time.[2]) According to Locke, I can know that I am the same person who performed a certain action if I have a clear memory of doing so. Yet I have no knowledge of whether I have, or am, a particular kind of substance; for all I know, there may have been a succession of substances associated with my body since the time of the action I recall performing. Indeed, if reincarnation occurs, I may have, or be, the same soul as someone living centuries ago or someone who will live centuries in the future. Yet, without any memory of such a past or anticipation of such a future, what possible self-regarding concern can I have for the relevant individuals' activities or welfare? Persons, we ordinarily think, are beings who can take responsibility for their actions and be concerned about their own future well-being.[3] The natural thesis, according to Locke, is that personal identity consists in "the same consciousness" – more precisely, continuity of a mental history over time, where present and past transient moments of awareness are connected by memory.[4]

Thus, without denying that we are (or are intimately associated with) substances, Locke held that sameness of substance is irrelevant to personal identity, which consists in *relations* among moments of consciousness. This thesis represented a radical break from tradition.[5] It vindicated nonskeptical attitudes about self-knowledge over time while accounting for ordinary notions of personal responsibility, prudential concern, and the like. (More conservatively, it also allowed for an afterlife.[6]) Locke's thesis provoked vigorous rebuttals from critics, but the criticisms need not

---

[2] Indeed, he writes that "the more probable opinion is, that this consciousness is annexed to, and the affection of, one individual immaterial substance" (*Essay Concerning Human Understanding*, 2nd ed. [1694], Bk. II, ch. 27, sect. 25).

[3] Ibid., ch. 27, sect. 26.

[4] Ibid., ch. 27, sects. 9, 10.

[5] For an illuminating discussion, see Raymond Martin, *Self-Concern: An Experiential Approach to What Matters in Survival* (Cambridge: Cambridge University Press, 1997), pp. 3–5.

[6] *Essay*, Bk. II, ch. 27, sect. 15.

detain us here.[7] The live issue is whether something along the lines of Locke's theory – some view stressing psychological continuity over time – will hold up under scrutiny. Derek Parfit answers affirmatively.

Parfit's Innovations

In recent decades, the old problem of personal identity has been given new life by myriad philosophers in the analytic tradition.[8] In the current discussion, the *psychological view* has a major, and probably dominant, place. The psychological view is the general approach taken by various theories that in some way understand personal identity in terms of psychological continuity. Amid the great diversity of theories within the psychological approach, the most influential contemporary representative is Parfit.[9] Nowadays, those who reject the psychological view – or the now mainstream way of developing it, which stresses *continuity of mental contents* – generally feel compelled to reply to Parfit, and the present author is no exception.[10]

Parfit's theory of personal identity is sophisticated and multifaceted, but our purposes require noting just a few refinements of the general Lockean approach. First, Parfit is explicitly *reductionist* about persons: "A person's existence just consists in the existence of a brain and body, and the occurrence of a series of interrelated physical and mental events."[11]

---

[7] Two prominent early critiques appeared in Joseph Butler, "Of Personal Identity" (First Appendix to *The Analogy of Religion* [1736]) and Thomas Reid, "Of Memory" (in *Essays on the Intellectual Powers of Man* [1785], chs. 4, 6). For a good discussion, see Harold Noonan, *Personal Identity* (London: Routledge, 1989), ch. 3.

[8] Twentieth-century contributions that helped to catalyze the current discussion include Sydney Shoemaker, *Self-Knowledge and Self-Identity* (Ithica, NY: Cornell University Press, 1963); David Wiggins, *Identity and Spatio-Temporal Continuity* (Oxford: Blackwell, 1967); Peter Geach, "Identity," *Review of Metaphysics* 21 (1967): 3–12; and Bernard Williams, "The Self and the Future," *Philosophical Review* 79 (1970): 161–80.

[9] An early twentieth-century effort to refine Locke's effort that did not generate much discussion at the time is H. P. Grice, "Personal Identity," *Mind* 50 (1941): 330–50. More distinctively contemporary representatives of the psychological approach are Sidney Shoemaker, "Persons and Their Pasts," *American Philosophical Quarterly* 7 (1970): 269–85; John Perry, "Can the Self Divide?" *Journal of Philosophy* 69 (1972): 463–88; David Lewis, "Survival and Identity," in Amelie Rorty (ed.), *The Identities of Persons* (Berkeley: University of California Press, 1976): 17–40; Derek Parfit, *Reasons and Persons* (Oxford: Clarendon, 1984); Noonan, *Personal Identity*; and Martin, *Self-Concern*.

[10] See, e.g., Mark Johnston, "Human Beings," *Journal of Philosophy* 84 (1987): 59–83; Christine Korsgaard, "Personal Identity and the Unity of Agency: A Kantian Response to Parfit," *Philosophy and Public Affairs* 18 (1989): 101–32; and Peter Unger, *Identity, Consciousness, and Value* (New York: Oxford University Press, 1990).

[11] *Reasons and Persons*, p. 211.

A person is not some separately existing entity – such as an immaterial soul – above and beyond the physical and mental elements; a complete description of the world could be given impersonally, that is, without mentioning persons.

Second, although persons apparently cannot exist without bodies, *personal identity over time depends on psychological relations.* So far, this point about identity is pure Locke. But Parfit's supporting reasoning addresses a question that may remain for someone who is sympathetic to a modern substance-based view: If we assume that the human body, or brain, is the relevant substance, what's wrong with holding that our identity consists in the continued existence of our body or brain?

Parfit argues that if physical continuity were essential to personal identity, it could not be continuity of the whole body but at most that of the brain; we can survive organ transplants and amputations, and seemingly replace or lose other body parts at least until we get to the functioning brain. As for the brain, we could lose some of it and still survive, so a defender of a material-substance view should insist only on the continued existence of enough of the brain to be the brain of a living person.[12] But the importance of the brain, Parfit continues, is derivative, stemming from what it *does*. A brain that no longer supported any form of psychological continuity would not preserve a person.[13]

In his view, whatever identity may factually depend on – in normal cases, the brain – what it *consists in* is, roughly, psychological continuity. Several definitions permit him to be more precise than Locke was about psychological continuity. First, *direct psychological connections* are such connections as that between a memory and the past experience remembered, between an intention and the later action that fulfills it, and the persistence over time of a desire, a belief, or another psychological feature.[14] Following are two important relations and a key concept: "*Psychological connectedness* is the holding of particular direct psychological connections" and "*Psychological continuity* is the holding of overlapping chains of *strong* connectedness," where *strong connectedness* obtains "if the number of direct connections, over any day, is *at least half* the number that hold, over every day, in the lives of nearly every actual person."[15] An important implication, addressing an objection to Locke's view, is that I can be the

---

[12] Ibid., p. 204.
[13] Ibid., p. 208.
[14] Ibid., p. 205. Elsewhere, he mentions the example of a character trait (see, e.g., p. 261).
[15] Ibid., p. 206.

person who committed a crime, even if I do not remember performing the deed, so long as there are sufficiently many overlapping chains of direct psychological connections between me now and the perpetrator of the earlier deed.

But identity does not consist only in psychological continuity and/or connectedness; it requires a "nonbranching" form – *uniqueness* – as well. If person *A* somehow "fissioned" into two persons, *B* and *C* – say, as a result of bisecting a living brain and successfully transplanting the two hemispheres into two different bodies – *A* could be psychologically connected with both *B* and *C*. But since *B* and *C* are distinct persons, having two different bodies and beginning two diverging life paths, *A* cannot be identical to both *B* and *C* (assuming identity is a transitive relation).[16] Identity is preserved only if there is a unique continuer of the original person's mental life.

Finally, for Parfit, another condition of personal identity is that continuity and/or connectedness have *the right kind of cause,* about which he distinguishes three possible views.[17] In the original 1984 printing of his book, he defended the "wide" view that *any* cause of continuity and/or connectedness counts for identity. This implies that, if my body were destroyed and elsewhere a single exact replica were created out of fresh material – even if the duplication were achieved by a flukelike causal connection rather than by a reliable means – I would be identical to my replica. In the second printing, however, Parfit withdrew his support for the wide view. To take a position on the question of what sorts of causes are "right," he stated, would conflict with his view that, because identity is not ultimately what matters in survival (as explained later), we should not try to decide among these three possible views.[18]

Parfit has attempted to analyze personal identity – to reduce it to more basic conceptual elements – by providing criteria that, in order to avoid circularity, do not invoke facts regarding a person's identity. He defines personal identity as Relation *R* (continuity and/or connectedness with

[16] An alternative, "four-dimensionalist" interpretation of fission cases is that persons *B* and *C* existed all along, sharing *A*'s body before fission took place. Just as a particular road can constitute one or two highways, depending on whether the road branches into two separate highways, there may be one or two persons associated with a particular human body, depending on whether fission occurs at a later time. See Lewis, "Survival and Identity."

[17] *Reasons and Persons,* pp. 207–8.

[18] See the note added in 1985 to "Introduction" (1986 printing of *Reasons and Persons*).

the right kind of cause) plus uniqueness.[19] In a radical innovation, he goes
on to argue that identity is not what primarily matters in survival; instead,
Relation *R* is what primarily matters.[20] But for now, we may focus on
identity, which adds uniqueness to Relation *R*, and his claim of avoiding
circularity – a claim that extends to Relation *R*.[21]

In advancing a charge of circularity, Joseph Butler targeted the psychological connection on which Locke places the most weight: memory.
While we surely remember only our own past experiences, we must distinguish genuine memories from mistaken or delusional pseudomemories.
Otherwise, a memory criterion of identity will imply that a lunatic who
wrongly thinks she remembers spearheading the White House initiative
for health care reform in 1993–4 really is Hilary Clinton. But how to
distinguish true memories from imposters, except by saying that true
memories are apparent memories in which the person remembering is
the person who actually had the earlier experience? Memory seems to
presuppose personal identity and therefore, argues Butler, cannot serve
as a criterion for identity.[22]

Parfit's strategy for avoiding circularity involves stipulating a definition for a concept broader than memory – *quasi-memory* – that does
not presuppose identity. I have an accurate quasi-memory of some experience if (1) I seem to remember having an experience, (2) *someone*
did have this experience, and (3) my apparent memory is causally dependent, in the right sort of way, on that past experience.[23] Ordinary
memories are accurate quasi-memories of our own past experiences.
We might even imagine cases of quasi-remembering others' experiences
through technologies that copy memory traces from one person's brain
to another's brain; as long as the subject did not assume that she had
the past experience she quasi-remembers, she would avoid delusion.
The important move is allowing the concept of causation to do the

---

[19] Ibid., p. 263.
[20] Ibid., chs. 12, 13. Equally controversially, he argues that any cause should qualify (pp. 215,
217). This does not contradict the post-1984 neutrality on what should count as the right
kind of cause, because that neutrality concerned the analysis of identity, not what matters
in survival.
[21] *Reasons and Persons*, p. 220. Hereafter, I will use *psychological continuity* as elliptical for
"psychological continuity and/or connectedness." Doing so is more convenient and
coincides with the usage of most of the relevant literature. And only in discussing Parfit's
view will I use the term *Relation R*, which, by including the requirement of the right kind
of cause, covers it up.
[22] "Of Personal Identity."
[23] *Reasons and Persons*, p. 220.

work Butler thought required an assumption of identity: distinguishing memories from delusional pseudomemories. (The lunatic's seeming to remember spearheading the health care initiative is not caused by actually having had the experience of doing so; and since her alleged memory isn't causally related to Clinton's earlier experience in any significant way – say, via a memory trace implant – she doesn't even quasi-remember the experience.) What the psychological theorist tried to say about memory can now be said in terms of quasi-memory, permitting a characterization of nondelusional memory without the circularity of referring to identity. Circularity is also avoidable with respect to other kinds of psychological connection by introducing the concepts of quasi-intention, quasi-belief, and so on.[24] If Parfit is right, this meets a major challenge to the general Lockean program.

With Parfit's theory – the most influential version of the psychological approach – in view, let us examine the strengths and weaknesses of this approach. After identifying some of its strengths and focusing on one in particular, I will argue that the psychological approach is subject to several damaging criticisms.

A Strong Connection to Some Everyday Practical Concerns

The psychological view has several strengths. First, as noted in discussing Locke, this approach coheres nicely with our assumption that we typically have knowledge about our own identities over time. For the major resources that make self-knowledge over time possible – memory and anticipation – are also, on this view, *constitutive* of our identity. By contrast, if our identity consisted in sameness of spiritual substance, knowledge of one's own identity would be mysterious.

Second, the psychological approach tracks most of our intuitions regarding identity – not only in the circumstances of actual human lives, as bodily-continuity theories may do about as well, but also in the sorts of imaginative hypothetical cases in which analytic philosophers specialize. Where consciousness parts ways with the body or even the soul, we tend to think that the person goes where consciousness goes. Locke's hypotheticals tend to convince us of his conclusion, as do the modern brain-transplant case of Shoemaker and its philosophical progeny.[25] Thus, if a

---

[24] Ibid., p. 261.

[25] Shoemaker, *Self-Knowledge and Self-Identity*, pp. 23–4. Ironically, Shoemaker initially shied away from the Lockean conclusion that others found so strongly supported by his thought experiment (ibid., p. 247).

prince and a cobbler suddenly acquired the memories, personality, and intentions previously associated with the other's body, we would tend to believe that somehow two persons had switched bodies. We would also tend to believe that if Tom's entire brain were transplanted from his original body to another body, maintaining psychological continuity, Tom would go with his brain and acquire a new body. Now, if we consult our reactions to a broad enough range of cases – and sometimes, as Williams brilliantly argues, if we consider the same case from different perspectives – we are likely to find some inconsistencies in our beliefs about personal identity.[26] But it seems fair to say that reactions *tend* to trace a Lockean path, explaining why the intuitive case method has served as the major argumentative tool for the psychological view. While it is difficult to deny that broad intuitive appeal is a theoretical strength, later we will question the reliability of the case method in connection with highly contrived, unrealistic scenarios.

This section will focus on the third general strength of the psychological approach: its apparently strong connection with certain practical concerns that are central to our lives. Personal identity is presupposed in many of our most basic practices and institutions. And the idea that identity involves some sort of psychological continuity accounts rather well for our beliefs about identity in these contexts. Let me elaborate.[27]

For starters, the idea that persons persist through time is assumed by our conception of persons as *moral agents*. Ordinarily, persons are responsible for their actions and can be held accountable for them. Unless emotionally very abnormal, persons can take pride in their good works and feel guilty, ashamed, or contrite when their behavior falls below social norms. Correspondingly, others may praise or criticize the agent, or reward or punish her.

These features of ordinary human life assume that persons persist through time – that is, maintain their identities. If persons were thought to exist only momentarily, any punishment for a past misdeed could elicit a legitimate protest: "You got the wrong guy!" But moral agency and therefore accountability arise *only with the somewhat developed form of consciousness that permits the individual to regard herself as existing over time;* having a body, for example, is insufficient. Memory is the most fundamental (though

---

[26] Williams, "The Self and the Future."

[27] The following reflections were stimulated in part by a discussion in Marya Schechtman, "The Same and the Same: Two Views of Psychological Continuity," *American Philosophical Quarterly* 31 (1994), pp. 200–1.

not the sole) cognitive resource for recognizing that one did something in the past and, a fortiori, that one did something right or wrong. In sum, psychological theories seem to make sense of such ideas as responsibility and remorse.

The psychological approach also connects rather plausibly with a more prospective aspect of persons: *their capacity to act intentionally, make life plans, and watch after their own interests.* Planning and prudence are central features of human life. But, to go to graduate school with the idea of later teaching, and to work hard in order to land a decent job, presuppose that one will be around later to carry out the plan and enjoy the fruits of one's earlier labor.

Now, caring about one's own future welfare, which makes prudence possible and planning sensible, arises out of a form of consciousness that embodies a sense of persisting over time. (Intending to do something or anticipating a future experience are rudimentary elements of such a consciousness.) Moreover, one might hold that, in acting and planning prudently, we not only assume that we will exist at a later time, but specifically we assume that we will be *conscious* later – a claim that fits especially well with the psychological view. Admittedly, this thesis might be debated, for instance by those who claim to have prudential concern about what would happen to them in PVS. Either way, the Lockean can justifiably assert that prudence and planning paradigmatically, if not exclusively, presuppose continuity of mental life between present agent and future person. No assumption about a persisting soul seems necessary. And sameness of (living) body or brain, even if factually necessary, seems less central than sameness of mental life to what we care about in self-concern – and, again, is arguably insufficient for prudential attitudes.

Consider now some general features of *social life among persons.* We relate to each other not as momentary beings but as individuals with ongoing histories. While this is true, to some extent, even of our interactions with strangers, it is especially important in the context of family ties, friendships, and other close relationships. Here we not only assume that others have personal narratives, which involves memories and plans, among other things (an assumption we make with strangers as well); we also take the narratives into account in our interactions with familiar individuals. And, in doing so, we attribute to each individual a personality and character, which involve relatively enduring traits. Such traits are psychological features of an individual. To have a cheerful disposition, or the virtue of honesty, for a long stretch of one's life is a kind of psychological connection, suggesting that psychological theories are on track here as well.

To a large extent, then, our social interactions and relationships assume that others have ongoing narratives and such psychological features as personality and character traits. At the same time, others' bodies also clearly play a major role in social life, especially in recognizing one another. However, in an unusual case featuring bodily continuity and psychological discontinuity – say, where injury or brainwashing results in extensive amnesia and a fundamentally new outlook and personality – many people may favor psychological criteria and regard the present individual as a numerically distinct person who must be related to as such. The more drastic the change, the more likely they would do so.

Finally, *belief in an afterlife* fits well with psychological theories. Even those who do not share this belief tend to agree (1) that the notion of an afterlife is perfectly coherent and (2) that continuation of one's mental life after one's death – with or without a body – *would* entail the person's survival (assuming there is no branching and the cause is acceptable). And while traditional thinkers might assume that a soul is a necessary vehicle for psychological continuity after death, sameness of mental life seems to be the most salient or decisive condition for personal survival. If a soul survived in an afterlife but carried no mental connection with life before death, there would seemingly be little justification for the claim that the person survived.

A parallel point applies to the notion of a *beforelife* and reincarnation generally. Even if a mystic convinced you that you are animated by the same soul that once animated Cleopatra, you would probably not believe that you are the same person as she. You would take that possibility seriously only if you had some apparent memories of her life, foreign leaders were uncannily attracted by your personal style, and the like.

In sum, then, the Lockean approach strongly connects with such major practices and institutions as taking and attributing responsibility, planning and prudence, and responding to others as persons. This approach also fits well with common thinking about an afterlife (and with less common thinking about reincarnation). In view of these strengths – along with the illumination of self-knowledge of identity and intuitive support from the case method – the general idea of understanding personal identity in terms of psychological continuity is prima facie promising. But let us turn now to concerns about the methodology from which this approach draws its major support.

CONCERNS ABOUT THE INTUITIVE CASE METHOD

The chief argumentative tool for the psychological view has been the intuitive case method. Using this method, we track our intuitive responses to various cases involving persons – cases that include the actual but prominently feature the very hypothetical, such as Locke's case of the prince and the cobbler – in an effort to produce generalizations about personal identity. But the intuitive case method has significant shortcomings.[28]

By considering a range of hypothetical cases that include brain transplants, apparent body switching, the gradual but complete replacement of natural body parts with bionic parts, "fission" of one person into two, "fusion" of two people into one, teletransportation, apparent transformation into a different sort of creature, and the like, the intuitive case method seeks only those conditions for the survival of persons that are *conceptually* necessary. Physical impossibilities are treated as irrelevant to the *bare concept* of persons (or subjects) and their identity.[29] Many thought experiments proceed from a first-person standpoint, from which it seems imaginable to undergo even Kafka-like changes of bodily form – changing into a cockroach, say, or even a teapot! – so long as psychological continuity is maintained. Note that from a third-person perspective, such a thought experiment is less convincing. If a person appeared to be spatiotemporally continuous with a cockroach or teapot, we might judge that the person had somehow been destroyed and *replaced* with a different sort of object, not that the person became a cockroach or teapot. But, one might reply, wouldn't this third-person description be fully convincing: Jane woke up, looked into the mirror, and saw (only) Tarzan's body? Yes, if we accept the assumption that *it was really Jane* who gazed into the mirror – and not, say, Tarzan in a state of profoundly psychotic identity confusion. But to accept that assumption is to beg the question of identity.

Besides, if we're imagining transformations from a first-person perspective, why should even the psychological continuity characteristic of persons be strictly necessary? So long as we carefully distinguish identity over time and *evidence* for identity, a distinction that makes sense of the

---

[28] For a formidable critique of Parfit's and Peter Unger's employment of the case method in support of their views *regarding what matters in survival,* see Martin, *Self-Concern,* pp. 21–7.

[29] This point is developed in Mark Johnston, "Human Beings," *Journal of Philosophy* 84 (1987), p. 60. Johnston's critique of the case method has significantly influenced the present discussion. See also Kathleen Wilkes, *Real People* (Oxford: Clarendon, 1988), ch. 2.

idea that we might persist even if massive psychological disruption prevented our knowing about our past, it seems possible to imagine surviving with psychological discontinuity. I can imagine surviving as a severely demented elderly man with so little psychological continuity with my past that I would not even qualify as a *person* (as defined in Chapter 1). I can imagine being such a subject even if I'm not confident about the details of what such an existence would be like. If I thought this subject would experience great pain, I would (self-interestedly) fear it.[30] Thus, *contrary to what proponents of the psychological view typically claim (as explained in greater detail in a later section), the intuitive case method suggests not that we are essentially persons – beings whose complex mental lives feature self-awareness over time or continuity of mental contents – but that we are essentially subjects or minds, beings with the capacity for consciousness.*[31]

Have I contradicted myself by using the very case method I criticize in suggesting that I could become an extremely demented old man? No, for I don't claim that the case method is useless, just that it is fallible. And it is considerably more fallible in farfetched cases like those featuring Kafka-like changes than in realistic cases like those involving dementia. Finally, my major claim is conditional: *If* one accepts and uses this method, this is where it really leads.

The first-person perspective that the case method often employs ensures only that a subject, as opposed to a psychologically continuous being or person, persists through the changes imagined "from the inside." But the bare concept of a persisting subject is extremely general and vague. Importantly, it doesn't include any persistence conditions that are *only factually necessary* – as opposed to logically or conceptually necessary.

This method risks erring in a way reminiscent of Descartes's classical mistake: taking the wide boundaries of what we can imagine ourselves to be – for Descartes, of a thing that thinks – to determine what we actually are. It is of more than historical interest that the psychological approach, at least as usually developed, comes within a metaphysical hair of substance dualism, the view that there are two mutually irreducible kinds of substances: material and immaterial. It is clear from Parfit's reductionism about persons and his rejection of dualism that in his view persons must be embodied, but the details of embodiment seem irrelevant. When he asserted that a person could survive as long as *something* caused continuity

---

[30] Jeff McMahan makes the same point ("The Metaphysics of Brain Death," *Bioethics* 9 [1995], p. 110).

[31] McMahan (ibid., p. 102) and Johnston ("Human Beings," p. 70) reach the same conclusion.

of mental life (since any cause would do), he left open the logical possibility that a person might end up having no brain at all. Later, he expressed neutrality about this wide view, declining to take a position on what sort of cause was necessary *for identity*. But since he still held that any cause would do for Relation *R* – which connects *person stages* even if a single person's identity isn't maintained – his view suggested no restrictions on what might cause a person's mental life. Thus, a person is just an embodied self-awareness. (Alternatively, we may say, considering where the case method really leads, that a *subject* is just an embodied locus of mental life.) The requirement of embodiment, motivated only by a background assumption of materialism, is all that stands between the psychological view and substance dualism.

The risk of employing such an unspecific conception of ourselves as a basis for understanding our identity is that we might, in hypothetical cases, overgeneralize from ordinary cases of identifying people over time.[32] Specifically, *we might illegitimately generalize from the fact that in everyday life psychological continuity provides sufficient grounds for asserting identity to the conclusion that psychological continuity provides such grounds in every imaginable case.* Additionally, what may only be normally sufficient *evidence* for asserting identity, the apparent continuation of a mental life, may wrongly be taken for *criteria* of identity.

Finally, it is worth noting that *the intuitive case method seems to generate logically contradictory results.* Consider two descriptions of a case presented by Bernard Williams.[33] The first description presents a scenario that intuitively induces us to say that two people switched bodies. Person *A* undergoes a scientific procedure that extracts all the information stored in his brain (without extracting the brain itself, let's say) while rendering him temporarily unconscious. Person *B* simultaneously undergoes the same procedure. Then the information extracted from *A*'s brain is "downloaded" into the *B*-body brain, and vice versa. After the *A*-body person and the *B*-body person (described this way to avoid begging the question as to their identity) regain consciousness, the *A*-body person speaks and acts as if he has *B*'s personality and character as well as detailed memories of *B*'s life, and no memories of *A*'s life, prior to the operation; meanwhile, the *B*-body person similarly seems to have acquired the mental life of *A*. This is essentially Locke's prince and cobbler example updated with a dash of science fiction to suggest that the apparent transfer of mental lives has an intelligible mechanism. Except perhaps where presented right after

---

[32] Johnston, "Human Beings," pp. 75, 80–1.
[33] "The Self and the Future."

the description of cases that highlight and seemingly recommend bodily criteria for identity,[34] this case consistently induces the intuition that *A* and *B* swapped bodies, following their minds. If it is stipulated that, following the operation, the *A*-body person will be tortured while the *B*-body person will receive a large sum of money, we will respond, in effect, "Good for *A*; tough for *B*."

But now consider this. You are told that you will soon be tortured. You're terrified. Your interlocutor adds that, before the torture, you will lose your memories. "Great," you think sarcastically, "massive amnesia followed by torture." When you hear that, in addition to losing your memories, you will acquire what seem to be memories of someone else's life, this seems even worse: "*Madness* followed by torture." Suppose now you are further informed that another person will lose his memories and then seem to gain the memories of your life before receiving a lot of money. This would hardly cheer you up; indeed, it might add a bit of envy to all the dread of your anticipated future. But, except for the question-begging use of pronouns – which imply that you will remain with your original body throughout the events described – this scenario is the same as the first one (which induced us to say that *A* and *B* swapped bodies), with you occupying the role of *A*. When the scenario is presented this way, it suggests that *A* and *B* *do not* swap bodies. Supporting this interpretation is this seemingly unimpeachable thesis: "one's fears can extend to future pain whatever psychological changes [such as the onset of amnesia or insanity] precede it."[35]

This interpretation is also consistent with my earlier claim that the intuitive case method really supports the view that we are bare subjects of mental lives, since *A* and *B* on this interpretation maintain their capacity for consciousness despite major psychological discontinuity. As Johnston puts it, "we easily understand the *stipulation* that A *undergoes* or survives throughout the 'reading into' his brain of B's psychology and then suffers severe pain while B *undergoes* the 'reading into' his brain of A's psychology."[36] But perhaps, in the end, it is most accurate to say that the case method *tends* to support the bare-subject view. After all, when Williams's case is presented in the first way, employing a third-person perspective, the body-swapping intuition remains strong. My suggestion is to take the apparently inconsistent – or, at best, uncertain and ambiguous – results

[34] Johnston notes this possible exception ("Human Beings," p. 81).
[35] Williams, "The Self and the Future," p. 180.
[36] "Human Beings," pp. 68–9.

of the intuitive case method to provide additional reason to regard it with suspicion. I conclude, then, that this method does not seem entirely reliable, especially regarding cases that are not known to be factually possible.[37]

WHY TAKE ESSENTIALISM SERIOUSLY?

Earlier I suggested that the issues of our numerical identity and our essence are closely linked. But why assume that we human persons have any essence at all? Couldn't we maintain that we have no essential properties?[38]

Consider the matter this way. Obviously, there are many kinds of things: plants and animals, boxes and books, planets and galaxies, houses and paintings. And presumably every particular thing is of at least one kind. But the particular things of the world, or at least most that interest us (maybe not all subatomic particles, for example), persist over time; they don't last just for a moment. At the same time, things are constantly changing, at least in subtle and microscopic ways. The fact that things can persist through time, despite changing, along with the fact that things can also go out of existence (as all finite things can), means that there must be criteria of identity. These criteria of identity determine those conditions under which a thing will persist, or continue to exist, and those conditions under which it will not. And a particular thing's criteria of identity, or persistence conditions, will be determined by what kind of thing it is. Thus, a flower's persistence conditions have to do with biology, whereas a box's persistence conditions presumably concern structure, function, and perhaps material composition, but not biology. Our criteria of identity will also depend on whatever kind of thing we are.

One might respond, however, that we human persons are of many kinds without any one kind representing our essence. For example, I am an animal, a person, a male, a father, and a scholar. So one might argue

---

[37] Robert Nozick argues that, rather than generating inconsistency or proving unreliable, the case method vindicates the *closest continuer* theory of identity, which suggests that a *full* description of what happens to the A-body and B-body persons vindicates the body-swapping interpretation (*Philosophical Explanations* [Cambridge, MA: Harvard University Press, 1981], ch. 1). But I do not think Nozick's approach succeeds. It doesn't square with our conclusion that the case method really supports the bare-subject view, and it is subject to a counterexample from Johnston ("Human Beings," pp. 67–8) that I find persuasive.

[38] For a nice critique of such antiessentialism, see Lynne Rudder Baker, *Persons and Bodies: A Constitution View* (Cambridge: Cambridge University Press, 2000), pp. 35–9.

that I have criteria of identity only qua a particular kind. Thus, though I am now a scholar, I am not essentially a scholar; for many years I was not one, and I could continue to exist without being a scholar if I dropped scholarly activities from my life. Similarly, although I am a father, I have not always been a father, so fatherhood is not essential to my existence.

But notice that the examples of existence qua scholar and existence qua father both assert that *I* existed before becoming a scholar and father. That implies that there is something more basic to me than these attributes. Under what conditions would we say that I no longer existed at all? Although this is not uncontroversial, I claim that I would not exist as a corpse, which would be my bodily remains but not me. While someone might disagree, holding that I would exist as a corpse, she would surely admit that I would no longer exist under some circumstances, such as the obliteration of the corpse into a million scattered pieces. After all, I'm not an infinite or eternal being.

Now the idea that *I* can survive some, but only some, transformations implies that there are criteria for my identity and that the basic kind that determines those criteria also determines my essence. In general, a particular thing's *basic kind* or *substance concept* tells us, in a fundamental way, what that thing is, and not just what it does, where it is, or some other inessential fact about it. (By contrast, a phase concept or phase sortal – such as *child, adult,* or *student* – tells us something inessential about a particular thing, designating a kind to which that thing can belong temporarily.) A substance concept determines the persistence conditions for all things of the same basic kind.[39] I will argue that we human persons are essentially (living) human animals.

Another example will help to clarify these ideas. Ariel is, among other things, a college student, a person, and a human animal. It is true that, necessarily, if someone is a college student, she must be enrolled at a college or university. But this type of necessity is only *de dicto* – a matter of linguistic meaning. She can persist *as a college student* only so long as she is enrolled in a college or university. But Ariel can continue to exist after college, because she is not essentially a college student. Now, Ariel is also a person. And, necessarily, if someone is a person (as discussed in Chapter 1), she must have the capacity for sufficiently complex forms of consciousness. But, in my view, the necessity of having such a capacity

---

[39] On substance concepts, I have benefitted from Olson, *The Human Animal,* pp. 27–8. Olson credits David Wiggins, *Sameness and Substance* (Cambridge, MA: Harvard University Press, 1980).

is only *de dicto*. Ariel can exist qua *person* only if she retains such mental life. But, if I am right, Ariel can exist, even as a nonperson, so long as the human animal that she is, survives – in which case continuing the life of a particular human animal is her *de re* persistence condition, and being a particular (living) human animal is her essence. In general, *X* is an essential property of a thing if that thing cannot exist without having property *X*. If property *X* is both necessary and sufficient for the thing's existence, then *X* is *the* essence of that thing.

Someone who is skeptical about any form of (*de re*) essentialism might object as follows: Why can't we allow that Ariel persists qua *college student* only under certain conditions, qua *person* under different conditions, and qua *human animal* under still different conditions – without claiming that any of these or other kinds that she instantiates represents her *de re* persistence conditions or essence? The answer is that we all believe that Ariel, that individual, will persist through some changes but not others. If Ariel had no essence, and were not a member of some basic kind, we should not judge that she ever goes out of existence. Even if her body were blown to smithereens, and the bits disintegrated over 200 years into separate molecules, we should say – if she had no essence – that she still existed qua molecules. (Here I assume there is no afterlife.) This, I take it, is not what we really believe.[40] So I assume that we human persons are most fundamentally of some kind of thing, defined by our essence.

ESSENCE-BASED CHALLENGES TO THE
PSYCHOLOGICAL VIEW

One strategy for challenging the psychological view focuses on its implications regarding our essence and the boundaries of our existence. Various philosophers – including Eric Olson, who has done the most to develop these criticisms, W. R. Carter, P. F. Snowdon, Peter van Inwagen, and (to a lesser extent) myself – have employed this strategy.[41] Before

---

[40] David Wasserman asked whether this example unfairly considers a situation in which someone's body is disintegrated into *dispersed* pieces, which hardly constitute a plausible candidate for any sort of entity. But this reply suggests the intuition that human persons are essentially intact physical objects (as, say, solar systems are not). Moreover, we would presumably agree that if Ariel were pulverized and her remains shaped into a brick, she would not survive.

[41] See, e.g., Eric Olson, "Was I Ever a Fetus?" *Philosophy and Phenomenological Research* LVII (1) (1997): 95–109 and *The Human Animal: Personal Identity without Psychology* (New York: Oxford University Press, 1997); W. R. Carter, "Do Zygotes Become People," *Mind* 91 (1982): 77–95; P. F. Snowdon, "Persons, Animals, and Ourselves," in

developing the present challenge, let us note what the psychological view suggests regarding our essence.

Typically, psychological theorists state their major thesis in roughly this form: *A person* at one time and *a person* at another time are one and the same person if and only if there is [some specified type of psychological continuity] between them. This formulation assumes that at both times in question some person exists. But might someone who is a person at one time be identical to a nonperson at a different time? The previous formulation leaves that open. But the psychological view, at least in most of its variants, apparently embraces a stronger thesis: A person at one time and *a being* at another time are one and the same if and only if there is [some specified type of psychological continuity] between them. But since the specified types of psychological continuity are supposed to be those characterizing persons, this stronger thesis implies that the being with which any person is ever identical is a person. In other words, on this view, any being that is ever a person cannot exist at any time without being a person at that time – the *de re* thesis that I call *person essentialism*.[42]

Is the psychological view really committed to person essentialism? In principle, a psychological theory might venture only the *de dicto* thesis that a person's continuing to exist *as a person* requires psychological continuity. Indeed, this was apparently the approach of Locke, who carefully distinguished personal identity, identity of "man" (the human animal), and identity of soul without taking a stand on the essence of human persons.[43] But most of the theories held by contemporary representatives of the psychological view strongly imply the *de re* thesis. Occasionally, the thesis is stated explicitly.[44] More commonly, these theorists do not distinguish (1) the conditions under which one persists *as a person* from (2) *our* persistence conditions or identity, what we may neutrally call *human*

Christopher Gill (ed.), *The Person and the Human Mind* (Oxford: Clarendon, 1990): 83–107; Peter van Inwagen, *Material Beings* (Ithica, NY: Cornell University Press, 1990); and David DeGrazia, "Persons, Organisms, and the Definition of Death: A Philosophical Critique of the Higher-Brain Approach," *Southern Journal of Philosophy* 37 (1999): 419–40 and "Advance Directives, Dementia, and 'the Someone Else Problem'," *Bioethics* 13 (1999): 373–91.

[42] See my "Advance Directives, Dementia, and 'the Someone Else Problem'," pp. 379–82; and Olson, "Reply to Lynne Rudder Baker," *Philosophy and Phenomenological Research* LIX (1) (1999), pp. 161–2.

[43] *Essay on Human Understanding*, Bk. II, ch. 27. See also Mark Thornton, "Same Human Being, Same Person?" *Philosophy* 66 (1991): 115–18.

[44] See, e.g., Nozick, *Philosophical Explanations*, pp. 78–9, though he refers to "selves" rather than "persons."

identity. In short, they strongly imply that our identity is purely a matter of persisting as persons – which is true only if we are essentially persons. Moreover, as Olson points out, "everyone, or nearly everyone, who accepts a version of the psychological-continuity theory thinks that I should necessarily cease to exist if my psychological contents and capacities were completely and irrevocably destroyed. But the *de dicto* principle implies no such thing."[45]

Let us consider, then, theories that define persons in terms of certain psychological capacities, unpack personal identity in terms of psychological continuity, and (explicitly or implicitly) embrace person essentialism – and let's reserve the term *Personalism* for such theories. Most contemporary theorists in the psychological camp embrace person essentialism and therefore accept Personalism. So what's wrong with this view? Consider five major objections.

First, there is the *"fetus problem."*[46] Since Personalism holds that only beings with psychological capacities are persons and that we are essentially persons, and since (to take a conservative example) four-month-old fetuses lack psychological capacities, this view implies that we were never four-month-old fetuses. But that arguably contradicts both common sense and embryology, according to which we human organisms develop as fetuses, are born, and continue to develop through infancy, childhood, and other stages of life.

Second, there is *the problem of explaining the relationship between you, the person, and the early human organism.*[47] If you are essentially a person, then you came into being when the relevant psychological capacities emerged – either late in fetal development or, more likely, during infancy. What happened to your fetal or fetal-cum-infantile predecessor? Did it die? This would be surprising, since we are not aware of any such death that regularly occurs early in human development. It's also hard to believe it merely disappeared without dying, since it is a kind of organism and organisms die when they go out of existence. Another possibility is that you, the person, now overlap or coincide with the human organism that preceded you. Because the organism is numerically distinct from you (as nothing can precede itself), there are now two distinct beings associated with your body. This too is very hard to believe.

---

[45] "Reply to Lynne Rudder Baker," p. 162. Recall Parfit's reasoning that, although one could survive as a functioning brain, the latter's importance is derivative because a brain that no longer sustained psychological continuity would not support a *person.*

[46] See Olson, "Was I Ever a Fetus?", esp. pp. 95–7, and *The Human Animal,* pp. 73–6.

[47] See Olson, *The Human Animal,* pp. 79–81; and Carter, "Do Zygotes Become People?"

Third, there is *the challenge of explaining the relationship between you and the permanently unconscious being that will succeed you if you enter a PVS before biological death occurs.*[48] PVS is a medical condition in which a functioning brainstem permits the human animal to breathe spontaneously, maintain a heartbeat, metabolize, and continue other major biological functions except for the capacity for consciousness. Because the latter capacity is gone, so is personhood. Thus, Personalism provokes the question of how the (living) human organism in PVS, your successor, originated.[49] It's very hard to believe it was conceived or otherwise biologically brought into being just when PVS set in. Might it have emerged without being conceived or otherwise biologically brought into being? But it is a human animal, a type of mammal, and mammals come into being biologically. Perhaps instead it came into being, long ago, as the fetus. But that would make it a numerically distinct being that currently overlaps you, sharing your matter. This is hard to fathom.

Fourth, there is *the problem of implying that we are not animals.*[50] Consider again PVS. The person is gone, but a human animal continues to live, spontaneously breathing and so on. But, if a person and an animal can come apart, the person cannot *be* the animal, for nothing can outlast itself. So, you, the person, are not the animal surviving in PVS. But certainly there is no more than one animal life associated with every human life. Thus, Personalism implies that we are not animals at all – apparently contradicting biological fact.[51]

Finally, *there is a problem about counting conscious beings.*[52] We have every empirical reason to think that "higher" animals such as dogs and bears are conscious when they are awake as opposed to sleeping. And, surely, if a nonhuman animal can be conscious, so can a human animal. Now, if I am essentially a person and therefore not an animal (as suggested in the previous paragraph), then there are now, as I write these words, two conscious beings – both of whom would be substances – sitting in my chair: I, the person, and the distinct human animal. That seems one too many.

---

[48] See, e.g., Olson, *The Human Animal*, pp. 88–9. Cf. my "Persons, Organisms, and the Definition of Death," pp. 424–5.

[49] Defenders of the higher-brain definition of death challenge existing legal standards of death and deny that PVS patients are alive. I respond to this approach in Chapter 4.

[50] Olson, "Was I Ever a Fetus?", p. 101 and *The Human Animal*, p. 94.

[51] The formulation of the second, third, and fourth objections draws significantly from my "Advance Directives, Dementia, and 'the Someone Else Problem'," pp. 385–6.

[52] See Carter, "Do Zygotes Become People?", p. 94; and Snowden, "Persons, Animals, and Ourselves," p. 94.

"Does it?" one might ask. "After all, in a sense there are lots of conscious beings sitting in your chair right now, including a father, an adult, and a professor. So I don't see why the two-beings implication is problematic, either here or in the context of the second and third objections raised earlier." But this rejoinder misses the point of saying that, if Personalism is true, there are for each of us two conscious *substances*. For presumably *animal, human animal,* or some other biological category is a substance concept or basic kind in the sense defined earlier. Surely a worm is most fundamentally a worm, an animal, or a member of some other biological kind. So the Personalist will admit that the human animal associated with you is essentially a human animal, animal, or the like. In insisting that *you* are essentially a person, he implies that there are two substances associated with your (entire) living body, two entities with distinct persistence conditions: you and the animal. This is peculiar.

Such challenges as these provide some motivation for rejecting the Personalist theses that we are essentially persons and that our identity is a function of psychological continuity. (The motivation is only preliminary, for as we will soon see, a version of Personalism has been advanced to respond to these and similar challenges.) An alternative view is that we are essentially (living) human animals or organisms – members of some biological kind, anyway – and that our identity, like that of any organism, is a function of biological life. In other words, for any living thing X, X considered at one time is identical to Y considered at another time if and only if X and Y have the same (biological) life. Other alternatives, which define our identity only partly in biological terms, are also possible. In any event, we may understand the arguments of this section as laying down a single monumental challenge to any form of Personalism: *to provide a plausible account of the relationship, in the case of any human person, between the person and the human animal.*

A STRATEGY FOR REPLYING TO THESE CHALLENGES: THE CONSTITUTION VIEW

One Personalist view has been developed (in part) to address the challenge of plausibly accounting for the person–human animal relationship: *the constitution view.*[53] The most fully developed and carefully defended

---

[53] For an early statement of this approach, see Sidney Shoemaker, "Personal Identity: A Materialist's Account," in Shoemaker and Richard Swinburne, *Personal Identity* (Oxford: Blackwell, 1984): 30–43.

version of this view is Lynne Rudder Baker's, which I will treat as representative.[54] The central claim of the constitution view is that human animals *constitute* persons. Before discussing constitution, let us see why Baker's theory counts as a version of Personalism. First, she defines persons in terms of psychological capacities: Persons are beings with the capacity for complex psychological properties that she calls the "first-person perspective."[55] She also unpacks personal identity (at least partly) in terms of psychological continuity: Person *X* at one time is the same person *Y* at another time if and only if *X* and *Y* have the same first-person perspective.[56] Finally, she explicitly embraces person essentialism: "That any person is essentially a person falls out of the idea of constitution."[57]

The idea of constitution is most helpfully clarified through examples. A hunk of bronze may constitute a particular statue without being identical to it; destroying the shape by melting will destroy the statue but not the hunk of bronze. A large number of threads (arranged in a certain way) constitute a flag but aren't identical to the flag; taking the threads apart would destroy the flag but not the threads. *Mutatis mutandis*, according to Baker, the human organism constitutes the person without being identical to her. The organism preceded the person because only when the former was sufficiently developed to sustain a first-person perspective did a person exist; and the organism might survive the person if the capacity for a first-person perspective is lost before biological death occurs. Conceivably, a person could outlast the organism if the replacement of organic with inorganic (say, bionic) body parts permitted the continuation of the relevant first-person perspective.[58] In any case, if *A* constitutes *B*, *A* and *B* have different persistence conditions – and are therefore not identical – yet they are intimately related so long as *A*

---

[54] See Baker, "What Am I?" *Philosophy and Phenomenological Research* LIX (1) (1999): 151–9 and *Persons and Bodies*. The present discussion of Baker's view draws significantly from my "Are We Essentially Persons? Olson, Baker, and a Reply," *Philosophical Forum* 33 (2002): 101–20.

[55] *Persons and Bodies*, p. 59.

[56] Ibid., p. 132. Distinguishing her view from what *she* calls *psychological-continuity theories*, which she criticizes (ibid., pp. 125–30), Baker might deny that she unpacks personal identity in terms of psychological continuity. But I include among psychological-continuity theories not only theories that understand personal identity in terms of continuity of mental contents (the target of Baker's critique) but also theories that understand identity in terms of *the continuation of basic psychological capacities* (see, e.g., Unger, *Identity, Consciousness, and Value*). Presumably, a first-person perspective involves one or more such psychological capacities.

[57] "What Am I?", p. 158; see also *Persons and Bodies*, pp. 5–6.

[58] *Persons and Bodies*, pp. 11, 109

constitutes *B*. Constitution, on this view, is a kind of unity that falls short of identity.

Unlike some authors,[59] I accept constitution as an intelligible relation. Moreover, I find that appeals to constitution genuinely illuminate the relationship between such pairs as a hunk of bronze and a particular statue, and a multiplicity of threads and a particular flag.[60] Whether a human animal and a person are related in this way, however, is another question.

Equipped with her theory of constitution, Baker addresses the arguments presented in the previous section. First, she contends that we are not identical to, but merely constituted by, organisms that were fetuses. This assertion does not contradict common sense or embryology, she contends, because common sense is insufficiently fine-grained to distinguish identity and constitution, while embryology neither includes the study of persons per se nor entails anything about the relationship between a fetus and a person.[61] Moreover, because we can speak meaningfully of an "is" of constitution, there *is* a sense in which Baker, the person, was a four-month-old fetus: Again, she is constituted by something (a human organism) that once was such a fetus.[62] But, *strictly speaking*, because she is essentially a person, she did not come into being until the organism acquired the capacity for a first-person perspective.

So what happened to the early human organism when the person emerged? Baker answers that the organism simply continued to develop after it came to constitute a person. The reason some philosophers find this problematic, she suggests, is that they assume that if *A* and *B* aren't identical, then they're wholly separate things. But if constitution is an intelligible relation, we may deny that the person and the human animal are identical without implying that they are wholly separate things.

The same reasoning may be used to explain the relationship between the person and the human organism that survives in PVS: The organism that constituted a person no longer does so in PVS even though the organism continues to live. This assertion, the argument goes, is no more problematic than asserting that the hunk of bronze that once constituted

---

[59] See, e.g., Olson, "Reply to Lynne Rudder Baker," p. 164; and Jeff McMahan, *The Ethics of Killing: Problems at the Margins of Life* (Oxford: Oxford University Press, 2002), pp. 89–92.

[60] Somewhat similarly, Aristotle distinguished the matter of a particular substance from the substance itself (which is a unity of form and matter). Perhaps we may say that the matter constitutes the substance.

[61] "What Am I?", p. 156.

[62] *Persons and Bodies*, pp. 115–16.

a statue no longer does, due to melting, although the hunk of bronze still exists.

Does Baker's view imply that we are not animals? She claims not: "We are constituted by human animals, and when we say truly that we are human animals, we are using 'is' in the sense of constitution."[63] On the other hand, "what I am most fundamentally is a human person; and a human person is a [person] constituted (at least initially) by a human organism."[64] Since we could conceivably transform into bionic (and therefore nonanimal) persons, we might persist without any longer being constituted by animals. It is quite clear, then, that on this view I am not *necessarily* an animal; my persistence conditions and those of human animals are different. Therefore, *strictly speaking* – that is, using the "is" of identity – I am not an animal. But, since we may say that, in a sense – using the "is" of constitution – we human persons are animals, that suffices, Baker suggests, to avoid absurdity.

Finally, the constitution view allows us to deny the alleged implication of person essentialism that there are now two conscious beings (substances) sitting in this chair: you, the person, and the human animal. Rather, as just explained, there is one human person. Both you and the human animal that constitutes you are that person, that conscious being.[65]

In my estimation, Baker goes some distance toward meeting the challenge to Personalism of accounting plausibly for the relationship between persons and human animals. I find it more commonsensical to hold that we were once four-month-old fetuses. But Baker presents a sense in which we were such fetuses, invoking the "is" of constitution, and she may be right that common sense is too coarse to distinguish the "is" of constitution from that of identity. And I have noticed that some philosophy students and teachers find it plausible to say that they did not exist until personhood emerged: "There was no *me* before that point." Baker's rejoinder makes it unclear whether there really is a fetus problem.

In a similar way, Baker's appeals to constitution strengthen what Personalism can say about the relationship between the person, on the one hand, and the organismic predecessor and successor, on the other. Even when the human organism constitutes a person, it seems fairly plausible to say that there is a single unified being – a human person – undercutting

---

[63] "What Am I?", p. 157.
[64] Ibid., p. 155.
[65] *Persons and Bodies*, p. 198.

the sorts of *reductio* arguments advanced against Personalism, which treat the person and the animal as wholly separate beings. True, the two are always distinguishable, even as the animal constitutes the person, but that does not seem to entail that the two are wholly separate any more than the distinguishability of the bronze and the statue entails that the bronze statue is made up of two wholly separate entities. Although I remain inclined to believe that we *were* fetuses and may someday *be* human animals in PVS – strictly speaking! – I find Baker's alternative account of these relationships almost as congenial and do not claim authority for intuitions that probably fall far short of universal assent.

Can appeals to constitution save her from the oddity of saying that, strictly speaking, we are not animals? I am more troubled by this implication of her view. I doubt that saying we are *constituted* by animals brings us close enough to our animal nature. And, if it doesn't, then the problem of counting conscious beings may remain: Like the person, the human animal now sitting in my chair – who, after all, has a functioning brain – is conscious. We will return to these issues. In any case, it seems fair to say that Baker has done much to resuscitate Personalism following the challenges presented in the previous section.

Before we turn to challenges to her view, further detail about Baker's conception of persons – as beings with the capacity for first-person perspectives – will be helpful. Underlying all forms of self-consciousness, she claims, is a first-person perspective, which distinguishes persons from other sentient beings: "[This is] a perspective from which one thinks of oneself as an individual facing a world, a subject distinct from everything else. All sentient beings are subjects of experience (i.e., are conscious), but not all sentient beings have first-person concepts of themselves."[66] While she does not attribute first-person perspectives to nonhuman animals, she attributes beliefs, desires, and (apparently) intentional actions to many higher animals:

> We attribute beliefs and desires (perhaps in the vocabulary of aversions, appetites, and learning states) to nonhuman animals, which seem to be reasoning from a certain perspective. For example, the dog digs there because he saw you bury the bone there, and he wants it. . . . Such explanations do not thereby attribute to the dog or the infant any concept of itself as itself.[67]

Such means–ends reasoning from a perspective manifest what she calls *weak* first-person phenomena. But this is not enough for personhood:

---

[66] Ibid., p. 60.
[67] Ibid., p. 61.

"merely having a perspective, or a subjective point of view, is not enough for *strong* first-person phenomena. One must also be able to conceive of oneself as having a perspective."[68]

After acknowledging that some nonhuman primates (at least chimpanzees) have self-recognition that falls somewhere between weak and strong first-person phenomena, she reserves the term "first-person *perspective*" for subjects who manifest the strong type.[69] And personhood, thus understood, proves enormously important, giving rise to moral agency, autonomy (which involves the ability to form attitudes about one's own desires), and a sense of the future.[70] Baker submits that this conception of personhood, along with the thesis that we are essentially persons, explains the traditional assumption that we have special moral status in comparison with nonhuman animals.[71]

CRITIQUE OF THE CONSTITUTION VIEW AS DEVELOPED BY BAKER

Despite contributing substantially to the viability of Personalism, Baker's position has several major problems that put its adequacy in doubt. Here I present two of those problems.[72]

The Newborn Problem

First, her view has a newborn problem, which is more serious than the previously discussed fetus problem. The newborn problem, which involves making sense of our relationship to human newborns, has both an ontological and a moral dimension. Concerning ontology, *Baker's person essentialism and her conception of personhood (persons as beings with the capacity for first-person perspectives) imply that you and I were never born – that is, did not exist at the time of birth.*[73] Earlier I noted that some students and teachers

---

[68] Ibid., p. 64 (emphasis mine).

[69] Ibid., pp. 62–4, 67.

[70] Ibid., ch. 6.

[71] "What Am I?", p. 159.

[72] These are presented in "Are We Essentially Persons?", where I also contend (pp. 111–15) that Baker's ontology is deeply problematic for placing excessive weight on the person–nonperson divide and claiming that it coincides almost perfectly with the human–nonhuman divide.

[73] I discuss this implication of Personalism in "Advance Directives, Dementia, and 'the Someone Else Problem'," pp. 384–5; and "Persons, Organisms, and the Definition of Death," p. 424.

of philosophy allow that they were never presentient fetuses. Fewer admit that they were not born. Yet Baker's view implies as much.

That is because no newborn – let's reserve the term for humans in the first week of postnatal life – has the capacity for a first-person perspective. No newborn achieves strong first-person phenomena, which involve not just having a perspective but conceiving of oneself as having one. As suggested by their behavior and their relatively undeveloped brains, newborns, though clearly sentient, are not conceptually sophisticated. (Remember: Baker is talking about a kind of self-consciousness that she thinks chimpanzees lack.)

Would it help to stress the *capacity* for first-person perspectives, claiming that the capacity precedes the manifestation of the relevant phenomena? Baker analyzes having the capacity for a first-person perspective as (1) having all the structural properties required for such a perspective and (2) either (a) having manifested such a perspective beforehand [which doesn't apply to newborns] or (b) being in an environment "conducive to the development and maintenance of a first-person perspective."[74] Because temporarily comatose individuals maintain the needed structural properties and satisfy disjunct (a), they qualify as persons. Apparently thinking of disjunct (b), Baker also claims normal human newborns are persons.

Baker is fudging here. Newborns' brains develop at a tremendous rate in infancy.[75] So, assuming that they don't develop a first-person perspective for at least several months (as suggested by their behavior and neurology[76]), we must assume that they do not have the relevant neurological structures at birth. In support of her rather implausible claim that newborns already have the relevant structures and capacities, Baker cites a single work, quoting a supposed authority, before concluding that "from birth, development of a first-person perspective is underway."[77]

---

[74] *Persons and Bodies*, p. 92.

[75] See, e.g., Joseph Volpe, *Neurology of the Newborn*, 4th ed. (Philadelphia: Saunders, 2001), ch. 2.

[76] A newborn's behavior is not sophisticated enough to suggest possession of a first-person perspective. Indeed, in the first week of postnatal life (the time period specified to define *newborn* for our purposes), a normal infant's behavior is not very different from that of an infant born with anencephaly, *a condition characterized by complete absence of the cerebral hemispheres and therefore of consciousness!* On the instinctive, brainstem-controlled behavior of anencephalics, see Volpe, *Neurology of the Newborn*, p. 7; and Bruce Berg, "Developmental Disorders of the Nervous System," in Berg (ed.), *Principles of Child Neurology* (New York: McGraw-Hill, 1996), p. 666.

[77] *Persons and Bodies*, p. 92, note 1.

Here Baker seems to waffle between claiming (1) that the newborn already has a first-person perspective and (2) that the newborn is in a process of developing such a perspective. Claim (1) has already been criticized. Claim (2) hints at the idea of *potential.* Well, certainly normal newborns have the potential to develop a first-person perspective (or, more precisely, the potential to constitute beings – persons – who have such a perspective). Then again, so do fetuses, even presentient fetuses, but Baker holds that because such fetuses lack psychological properties altogether, they are clearly not persons.[78] Moreover, the term *capacity* suggests some *current* capability and so must be distinguished from potential. So Baker has not overturned the strong commonsensical (and scientific) presumption that newborns neither manifest nor have the capacity for first-person perspectives, though they certainly have the potential to develop this capacity (or come to constitute beings who have this capacity) within months.

Thus, because newborns are not persons by her criteria, Baker's person essentialism implies that, strictly speaking, none of us was ever born – that is, none of us existed at the time of birth. She can, of course, allow that *in a sense* you were born: You are constituted by a human organism that was born. But it would represent at least a modest advantage of any view that it could say, without qualification, that each of us was born.

Moreover, the newborn problem has a moral dimension: *Baker's view apparently implies that newborn humans have radically inferior moral status.* As just argued, newborns do not qualify as persons. But Baker accepts the traditional Western assumption (which I reject) that persons have radically superior moral status.[79] It is personhood itself, in her view, that underwrites significant moral status.

One who largely accepts the traditional view of moral status, which ascribes radically inferior moral status to nonpersons, might respond that Baker's view has acceptable implications for human newborns, who are *on their way* to becoming persons and are thereby distinguished from nonhuman animals, who never develop into persons. Perhaps this confers full moral status on newborns even if they are not yet persons. But this move assumes that *potential* personhood (or potential to constitute a person) confers the moral status associated with personhood on the traditional view, implying that even early fetuses have full moral status. Of course, some find it plausible to hold that early fetuses have full moral

[78] Ibid., p. 132, note 27.
[79] Ibid., p. 121.

status. For those who do not find this plausible, however, potential (to constitute) personhood does not suffice for moral status – in which case the newborn problem has a moral dimension for anyone who, like Baker, ties significant moral status to personhood. (Later in this chapter we will consider a modification of Baker's view that drops the requirement of personhood for significant moral status. In Chapter 7 we will return to the moral status of fetuses.)

This moral problem is intensified in the case of newborns who, although sentient, lack even the potential to constitute persons due to extreme mental incapacity. True, these humans will never have moral responsibilities. But I find it implausible that they have little or no moral status, as implied by the thesis that significant moral status requires the capacity for a first-person perspective.[80]

A Problematic View of Personal Identity

Another major difficulty with Baker's constitution view is an untenable view of personal identity. For Baker, person A at one time is identical to person B at another time if and only if A and B have the same first-person perspective.[81] The only material constraint on personal identity is the necessity of embodiment: "although human persons are not essentially human (they may [come to] have inorganic bodies), anything that begins existence as a human person is essentially embodied."[82] Yet the criterion of sameness of first-person perspective (in the absence of further material constraints) does not square well with several of her theses about personal identity.

Consider, first, certain thought experiments. Baker holds that two people could switch bodies, going where their mental lives apparently go:

I find the traditional thought experiments about bodily transfer – for example, the Prince and the Cobbler – utterly convincing when considered from a first-person point of view. Suppose that I wake up and look in the mirror and see a strange new body. What makes this person me, no matter what body constitutes her (or him!), is that she has my first-person perspective.[83]

---

80 Elsewhere I have argued that sentience confers significant moral status (*Taking Animals Seriously: Mental Life and Moral Status* [Cambridge: Cambridge University Press, 1996], ch. 3).

81 *Persons and Bodies*, p. 132.

82 Ibid., p. 214.

83 Ibid., p. 141.

She also finds intelligible these scenarios: (1) being teletransported, as one's original body is destroyed, to a new body in another location and (2) being transformed (Kafka-like) such that one ends up with a cockroach's body.[84] In each case, a persisting first-person perspective plus embodiment enables the person to continue.

If personal identity is determined by sameness of first-person perspective, *in what does a particular first-person perspective consist?* Baker cannot appeal to sameness of soul or immaterial substance, because she disavows substance dualism. According to Baker, personal identity cannot be analyzed reductively into nonpersonal terms; sameness of first-person perspective is as far as the analysis can go. But, then, how can we distinguish *conceptually* between these distinct scenarios: (i) The prince and the cobbler switch bodies, following their mental lives, and (ii) the prince and the cobbler stay with their bodies but suddenly become massively deluded, the prince acquiring the same mental contents and dispositions as the cobbler used to have, and vice versa? Presumably, Baker would respond as follows: All we can say is that, in the first scenario, the two first-person perspectives switch bodies, whereas in the second, the first-person perspectives remain with the same bodies but suddenly undergo a massive change of contents and dispositions. As indicated later, I doubt that this response will prove adequate.

Consider now her understanding of fission cases. If the mental contents of person *A* are transferred into the previously "blank slate" bodies *B* and *C*, as body *A* is destroyed, what happens to the original person? Perhaps the most common contemporary response is this: "*B* and *C*, who have different bodies and can go on to lead very different lives, cannot be identical. So *A* cannot be identical to *both B* and *C* (assuming transitivity of identity); otherwise, *A = B* and *A = C*, implying that *B = C*, contrary to our supposition. Nor can *A* equal *either B* (but not *C*) or *C* (but not *B*), because *B* and *C* are both psychologically continuous with *A* (and there's no reason to suppose that the mental states of only one of *B* or *C* bear the right sort of causal relation to the mental states of *A*). Therefore, fission causes *A* to go out of existence, despite her mental life continuing in two people."

While accepting that *A* cannot be both *B* and *C*, Baker denies that *A* necessarily goes out of existence. For she rejects the reduction of personal identity to psychological continuity *understood in terms of experiential*

[84] Ibid., p. 123.

connections. Rather, *A* continues only if either *B* or *C* has the same first-person perspective as *A*, in which case *A* is identical to whoever has that perspective. It does not matter that, from a *third*-person perspective, we could not tell which of *B* or *C* was *A*; it would be clear enough, Baker thinks, to *A* herself. But couldn't *both B* and *C* have *A*'s first-person perspective? Baker answers negatively, holding that a particular first-person perspective is necessarily unique, underscoring the idea that such a perspective cannot be identified by its contents alone (since contents can be duplicated).[85]

Baker's remarks on fission return us to this question: In what does a particular first-person perspective consist? How can we distinguish, *conceptually*, between (i) *A*'s continuing as *B* and (ii) *A*'s continuing as *C*? Baker, of course, would reply that in one case *A*'s first-person perspective continues in *B* and in the other case it continues in *C*. But what, precisely, *is* a particular first-person perspective? Again, she can't say it's a soul. Nor does she think it's a particular living body, brain, or brain part, because she rejects all biological and bodily-continuity views of our identity.[86] It seems to me that a particular first-person perspective, on Baker's view, must either turn out to be (1) some sort of "container" of experiences or (2) a mental life defined by its contents. But (2) is the sort of thing that can divide, because mental contents can be duplicated. Later we will return to the idea of a container.

For reasons similar to her grounds for rejecting common interpretations of fission cases, Baker rejects the thesis that personal identity is indeterminate. Parfit famously argued that reductionism – the claim that "[a] person's existence just consists in the existence of a brain and body, and the occurrence of a series of interrelated physical and mental events" – implies that identity is indeterminate. For continuity, whether psychological or physical, is a matter of degree and therefore admits of borderline cases in which it would be arbitrary to insist that identity either is or is not maintained; just as it is implausible that one grain of sand could make the difference between there being a pile or not, it is implausible that one more surviving cell or psychological connection could make the difference between a person's continuing to exist or not. The only way to avoid the conclusion of indeterminacy, Parfit argues, is to hold that we

[85] Ibid., p. 133.
[86] Ibid., pp. 119–25.

are "separately existing entities" like immaterial souls.[87] Baker claims that her theory can avoid indeterminacy without positing "separately existing entities."[88] I doubt it.

In arguing that personal identity is determinate, Baker states that "[e]ither I will experience waking up after the operation [in which various physical and/or psychological changes are induced] or I will not. From my own first-person point of view, there can be no indeterminacy."[89] Well, maybe it cannot *seem* to whoever wakes up that there is indeterminacy, but why can't there *be* indeterminacy? In response to the fission problem and the indeterminacy thesis, Baker says this:

> The following seem to me to be incontrovertible facts, easily discernible from a first-person perspective: Every morning when I wake up, I know that I am still existing – without consulting my mirror, my memory, or anything else. I can tell. . . . [I]f it is I, then I know without being told that I am the subject of that experience. If I have such an experience after fissioning, then I survive, constituted by whatever body I find myself related to via a first-person relation.[90]

Surely, at any moment I know that I am I, where both occurrences of the pronoun refer to the *present* subject. The assertion is safe because it is trivial. But Baker is interested in the nontrivial assertion that I (the presently existing subject) am identical to some "I" from the past – and know this.

Baker is right that, *in the actual world as we currently know it*, every morning I know that I am identical to a particular person who went to sleep several hours earlier. Such self-reidentification in actual cases typically furnishes knowledge of one's own identity. But it does not seem *necessarily* discernible from a first-person perspective that I am identical to some person from the past; massive delusion about identity is possible. For, again, these two scenarios are distinguishable: (i) I correctly remember the experiences of someone from the past (myself) and identify myself with that person; and (ii) I incorrectly think I remember the experiences of someone from the past (a deceased relative, say) and incorrectly identify myself with that person. Nor is scenario (ii) purely hypothetical, as cases of such profound psychosis are real occurrences.

Baker is really making two claims: (1) Personal identity consists in sameness of first-person perspective and (2) any person can know that

---

[87] *Reasons and Persons*, ch. 11.
[88] *Persons and Bodies*, p. 134, note 28.
[89] Ibid., p. 135.
[90] Ibid., p. 136.

she is identical to some person from the recent past. The possibility of massive delusion knocks down the epistemological second claim, assuming it is intended to apply to all possible – or even all real, including highly unusual – contexts. The possibility of massive delusion also raises doubts about the conceptual-metaphysical first claim. For the only difference between the two scenarios is that, in (i), I correctly identify whose first-person perspective I continue to have, whereas in (ii), I don't. The problem is that it remains mysterious what a particular first-person perspective is. Baker will not identify such a perspective with any physical substance, yet she also claims it is determinate and indivisible, precluding its being definable purely by its contents. All I can think of is that a particular first-person perspective is something like a container of experiences – or, if that metaphor is too crude (for suggesting that experiences are things that can be put into something else), a subject that cannot be identified with any physical object. That sounds like what Parfit calls a "separately existing entity," despite Baker's insistence to the contrary – and *that* sounds an awful lot like a soul.

Thus, Baker's theory of personal identity, when examined closely, either embraces substance dualism or remains shrouded in mystery. If the former, that would defeat her ambition of providing an alternative to standard views of identity without embracing substance dualism and would expose her view to all the difficulties of that theory. If the latter, much more needs to be said to explain the view and display its coherence. Either way, Baker has yet to produce a viable theory of personal identity.

Conclusion about Baker's Constitution View

In conclusion, Baker's version of the constitution approach seems inadequate. First, I doubt that her theory adequately handles the oddity of asserting that, strictly speaking, we are not animals. I suggest that we are, in the strictest possible sense, animals – not merely constituted by them – even if some of our more interesting characteristics are connected with our psychological capacities, including (but not limited to) those associated with personhood. And the thesis that we are identical to animals, I further suggest, is the simplest and best explanation for the intuition that there is a single conscious being in my chair. Baker's view is threatened by further difficulties as well, the two discussed here being the newborn problem, with its ontological and moral dimensions, and a problematic view of personal identity, which appears to steer uncomfortably close to substance dualism. Let me add my contention that

Baker's theory has problematic motivations. Besides relying too heavily on the intuitive case method, Baker romanticizes personhood. That is, she exaggerates its importance in terms of determining moral status, dividing up the world's creatures (since she thinks the set of persons equals the set of normal postnatal human beings), and defining what we are.

If my critique is fair, Baker's constitution view is not an adequate view of human identity and therefore doesn't rescue Personalism. But might some other version of the constitution view – some theory that maintains that we are essentially persons and constituted by human animals – prove adequate, rescuing Personalism? One possibility would be to abandon the claim that significant moral status requires personhood and therefore the capacity for a first-person perspective. We might hold that sentience confers significant moral status, while personhood is required for certain moral purposes, such as bearing responsibility for one's behavior, and perhaps slightly increases one's moral status. Such a revised view would avoid the newborn problem in its moral dimension.

The newborn problem would persist in its ontological dimension, however. Moreover, the inadequacy of sameness of first-person perspective as the criterion of personal identity would remain. Not only does this criterion have the problems indicated in the preceding section; earlier in this chapter we found that the (somewhat unreliable) intuitive case method apparently supports the bare-subject view. And, of course, *any* constitution view would imply that, strictly speaking, we aren't animals. In conclusion, I submit that no theory that understands our identity in terms of psychological continuity – either of mental contents or of basic capacities – while embracing person essentialism (where personhood is defined in psychological terms) will prove adequate.

THE BIOLOGICAL APPROACH: MOTIVATION, CONCERNS, RESPONSES

A Presumption Favoring a Biological View

Let's take stock. We examined the psychological view, in both classical and contemporary versions, and noted its apparently strong connection with certain practical interests. Then we found that the intuitive case method, on which this approach relies heavily, is far from infallible, especially with very farfetched cases. We proceeded to an argument that we should accept some form of essentialism regarding ourselves, leaving open the question of what we essentially are. After determining that the mainstream version of the psychological approach, Personalism, implies that human persons are essentially persons, we reviewed several objections to Personalism. Together, these objections established a monumental challenge to any version of the psychological view: to explain plausibly the person–human animal relationship. So we considered whether the constitution view, developed most fully by Baker, could meet this challenge and found that it could not. We concluded that any Personalist approach – any theory, that is, that understands our identity in psychological terms and accepts person essentialism – will prove highly problematic.

Might a psychological view that rejects person essentialism fare better? As we have seen, any plausible view will need to embrace some form of essentialism. So, if not persons, then what are we essentially? (Later, we will explore the possibility that we are essentially bare subjects or minds.)

At this point, the psychological approach appears somewhat less promising than one might have supposed in view of its recent dominance. Let us now introduce the possibility that a very different approach, the biological view, will prove superior – more plausible metaphysically, more coherent, and more in keeping with educated common sense.

Before describing this alternative view, however, I should note that I will beg one major philosophical question by assuming a naturalistic (materialist) perspective. In other words, I will assume that centuries of scientific illumination and philosophical reflection have provided ample reason to reject not just idealism, the thesis (which virtually no one holds anymore) that all substances are immaterial, but also substance dualism.[91] Fortunately, the contemporary philosophers whose views I challenge in

---

[91] Let me briefly mention three difficulties for substance dualism. First, reidentifying people becomes epistemologically problematic when common sense suggests that it should not be. For, if substance dualism is true, how could we know that the immaterial substance associated with a particular human being has persisted? Second, it is unclear how two radically different types of substance, material and immaterial, could causally interact, yet mental phenomena and physical phenomena appear to do so. Third, there is compelling behavioral evidence to suggest that split-brain patients – in whom the corpus callosum (which forms a neural bridge between the cerebrum's two hemispheres) has been severed as a treatment for severe epilepsy – have, as a result of the surgery, two centers of consciousness (see R. W. Sperry, "Hemisphere Deconnection and Unity in Conscious Awareness," *American Psychologist* 23 [1968]: 723–33). Substance dualism implies that one's mental life is rooted in one's soul or immaterial substance. Now, there is no reason to think that splitting a cerebrum would cause a soul to divide into two, yet dualism requires two souls to explain two centers of consciousness. While substance dualism faces these and other difficulties, naturalism is supported by the apparent dependence

this discussion also reject substance dualism, so I will be arguing from common ground in assuming naturalism. Anyone who accepts or remains neutral about substance dualism may take my thesis as conditional: *If* one accepts naturalism, *then* the biological view is the most sensible approach.

The assumption of naturalism has an important metaphysical consequence, namely, that we cannot exist as disembodied spirits because there is no such thing as an immaterial substance. Further, we know that after people die, their remains (whether in the form of corpses or ashes) are inanimate. Since there is no good reason to suppose that, for each of us, a second body is created and ample reason to deny this (e.g., the conservation of matter), a reasonable person who assumes naturalism would presumably hold that there is no afterlife of any kind. After dying, we live on neither in immaterial nor in material form; we don't live on at all. This consequence is important to the question of our essence. For it rules out the argument that, since we can *imagine* a nonbodily afterlife in which we continue the mental life we had while alive, we are essentially persons, conscious beings, or the like. What we can imagine ourselves becoming is irrelevant if the thought experiment is not merely technically but also metaphysically impossible. Our question is what we *actually* (that is, in the actual world) most fundamentally are: What is our basic kind? Thus, in the search for an answer, our imaginative investigations should take place within the constraints of what we know about the world. Let us proceed now to the major alternative to the psychological view.

According to the biological approach, we human persons – and, for that matter, human beings who are not persons – are essentially human *animals*, members of the species *Homo sapiens*. Human animals do not *constitute* us; as reasonably well-educated people assume, we *are* human animals. Therefore, on this view, there was a time when we who are now persons were not persons (namely, before the human animal developed the capacities that constitute personhood), and there might be a time in the future when we are no longer persons (say, if severe dementia reduces us to barely sentient beings).

The important claim here is not species membership per se, but membership in some biological kind.[92] Perhaps we are not essentially *Homo sapiens*, but essentially hominids or members of some other biological kind. Perhaps with enough genetic manipulations I could become a member of another

---

of all mental life on the operations of nervous systems, which are physical phenomena, and the absence of evidence for any irreducibly mental entities.

[92] Cf. Olson, *The Human Animal*, ch. 6.

hominid species (if species membership is determined by one's genetic makeup at a given time rather than by the species of one's parents). What's crucial is that we are essentially animals of *some* kind; we could not transform into plants or tables. While this statement of the biological view is somewhat vague and open-textured, it permits a clear contrast with psychological theories and it will prove to have important consequences. For the purposes of highlighting the contrast with psychological theories, I will continue to say that on the biological view we are essentially human animals, although I will not advance – and do not need – a thesis about the precise boundaries of our basic kind.

It follows from the biological view that we human persons are of the same basic kind as those unusual members of our species who never develop into persons. Again, a basic kind is the set of beings falling under a particular substance concept, a concept that determines the essence of beings to which it applies. (For example, I contend that *human animal* is our substance concept, whereas for Baker, *person* is our substance concept.) Even if our basic kind is not specifically *Homo sapiens*, other members of this species are of the same kind as we. For example, anencephalic infants, born without cerebra and therefore without the capacity for consciousness, are human animals.

On the present view, a species – or perhaps some broader biological classification – is a much more plausible candidate for a basic kind, or substance concept, than is a category like personhood.[93] After all, possible persons include not only normal postinfancy human beings but also certain extraterrestrial beings (I assume, the universe being enormous). If it turns out that inorganic materials can produce consciousness, then future supercomputers could also be persons. But what reason is there to suppose that all these beings would be of the same basic kind? Even if our basic kind is an extremely broad biological category that includes extraterrestrial creatures, supercomputers aren't even organic. Indeed, the *concept* of personhood permits us to imagine – contrary to naturalism – that there could be immaterial persons. If we include them among conceivable persons, it becomes even more difficult to maintain that they would all instantiate one basic kind. Rather than regarding personhood as a basic kind, we should regard it as comprising a set of capacities that things of different basic kinds might achieve.[94] On this view, personhood represents merely a phase of our existence.

---

[93] Cf. Peter French, "Kinds and Persons," *Philosophy and Phenomenological Research* 44 (1983): 241–54.

[94] This discussion of basic kinds was influenced by Olson, *The Human Animal*, pp. 31–7.

In addition to offering this naturalistic picture of what we are, the biological view claims a naturalistic, reality-based understanding of reidentifying individuals over time. If our identity consisted primarily in psychological phenomena, there is a strong case that it would consist in continuation of the capacity for consciousness – as in the bare-subject view. Now consider an everyday instance of recognizing someone. A man goes to bed, a man wakes up from that bed seven hours later, and we assume that the latter man is the former. On what basis? Observed continuity of bodily life. Nor do we take our belief in a single man's identity through the night as a good guess or just probably true; in ordinary cases, we take the belief as *knowledge*. But on the view that we are essentially bare subjects, we cannot know that the man persisted through the night, because we are in no position to rule out the possibility that more than one bare subject occupied his body successively during this time.[95] I take it that we do ordinarily have knowledge about the identity of others familiar to us, so I think the present criticism suggests that the bare-subject view is as problematic as the classical soul hypothesis of which it is the contemporary secular residue. So long as a bare subject is not defined in physical terms, ordinary cases of knowing another's identity after she has been unconscious will invite skepticism where none seems warranted.

Before we turn to concerns about the biological approach, let us make several observations about cases that are commonly thought to support the psychological view. Several of these cases provoke intuitive responses that seem to reify our minds. For example, the case of the prince and the cobbler typically provokes the reaction that the two individuals switched bodies, following their minds. *Following their minds?!* What sort of thing is a mind that it can jump from body to body? Minds, conceptually, are whatever has mental states. In the real world as we know it thus far, to refer to *minds* is to refer to brains functioning in certain ways. And it's not part of the prince and cobbler story that the two bodies swap brains. *Perhaps* in the future we will have good grounds to assert that minds include certain artificially made information-processing systems – supercomputers – made out of silicon and metal. Even if that proves possible, it will remain true that minds are certain highly complex physical objects whose characteristic functioning sustains consciousness. It is probably a bit misleading, because reifying, to refer to minds at all, inasmuch as doing so tempts us to think of minds as separable from the organs

---

[95] Johnston makes this argument ("Human Beings," pp. 72–3).

(or machines) that sustain consciousness, leading many self-avowed physicalists to sound like substance dualists.[96] And, again, from the present naturalistic perspective, it is assumed that substance dualism is false.

Thus, while the prince and the cobbler may behave as Locke describes them, if we assume no brains (or cerebra) are exchanged, then the psychological theorists' assertion that the two persons switched bodies, following their minds, is of dubious intelligibility – given the world as we know it. Since minds are physical objects in certain functional states, and no physical objects are exchanged between the prince's and cobbler's bodies, the two cannot "follow their minds" from one body to another.

Another reification-inviting case is that of teletransportation. After the precise location and type of every cell in my body is recorded, my body is destroyed, soon after which a signal is sent to Mars, where a cell-to-cell replica of my body is created. The person who now exists on Mars seems to remember my preteletransportation life and in other ways maintains psychological continuity with me; indeed, he believes, at least at a phenomenological level (whether or not he also believes it theoretically), that he is I.[97] A common response to this case is to say that teletransportation is a way to travel: One goes where one's mind goes. But, again, a mind is not simply a set of mental contents; nor is it a nonphysical container of such contents. A mind is a functioning brain (or other complex physical entity), and it is clear that the person on Mars has a numerically distinct brain from mine, which was destroyed. On the present view, then, teletransportation describes a situation of killing one person and creating a replica with identity confusion. Teletransportation is no way to travel.

Several Concerns about This Approach

While possessing many strengths, the biological approach also faces several important challenges, which I present and discuss in turn.

1. The Transplant Intuition. Suppose your brain is removed from your body and successfully transplanted into a different debrained human

---

[96] This point is made in Johnston, "Human Beings," pp. 79–80. See also McMahan, "The Metaphysics of Brain Death," pp. 102–3.

[97] A recurring theme in Martin, *Self-Concern*, is the possible divergence between our experiential beliefs and our theoretical beliefs about our identity and nature.

body, so that afterward the person who has that body is psychologically continuous with you prior to surgery. This scenario is not entirely far-fetched. Considering recent developments involving cloning, which was long commonly assumed to be impossible, I would certainly not rule out the possibility of brain transplants in a few decades' time. (By comparison, teletransportation – which involves recording every cell, including every neuron, in one's body and then producing a precise cell-to-cell replica – is far less likely ever to be achievable.) More importantly, this case does not dissociate one's mental life from one's brain. The *brain*, the organ of mentation, travels. So, unlike the cases of the prince and the cobbler and of teletransportation, the brain-transplant case really does feature a mind that leaves (most of) its original body. And the intuition, of course, is that you would go where your mind – your brain – goes and occupy a different body. Nor is the intuition different if we stipulate that it is not the *whole brain* (including brainstem) that gets transplanted, but only, as with Penelope in the case that opened this chapter, the *cerebrum*, the part of the brain whose functioning is necessary for consciousness and where the contents of consciousness are encoded.[98]

How does the brain-transplant intuition challenge the biological view? Consider, first, the variant that features just a cerebrum transplant. In the case involving Penelope, her original body retains its brainstem and continues most major bodily functions – including digestion, metabolism, circulation, and respiration – for a month following the transplant. That spontaneously breathing body seems no less alive than a presentient fetus, although neither manifests consciousness. But, if Penelope is essentially a human animal, she would seem to be the human animal that now lacks a cerebrum. Otherwise, we'd have to say that an entirely new human animal popped into being when Penelope left with her cerebrum. Now, if Penelope is that mindlessly breathing body, then she's not the person who seems to possess Penelope's mental life following the transplant. That's counterintuitive.

In a cerebrum-transplant case, the biological view denies that you go with your mind – in a perfectly responsible, nonreifying sense of *mind.* It may be less clear, initially, that the biological view delivers a similarly counterintuitive verdict in the whole-brain-transplant case. Suppose that Penelope's entire brain, including the brainstem, were transplanted

---

[98] Cerebral functioning is insufficient for consciousness. Also necessary is the activity of the ascending reticular activating system, or reticular formation, which serves as a kind of on/off switch. When the reticular formation is working, the contents of mental life encoded in the cerebrum can be consciously experienced.

while her original body (minus the brain) was kept breathing, circulating, and so on with artificial life supports. (An intact brainstem permits *spontaneous* cardiovascular functioning.) What does the biological view say about Penelope's fate in this case?

That depends on whether a body that maintains circulation, respiration, and other vital functions only by way of mechanical assistance is a living animal. In Chapter 4, I will argue affirmatively. However, some will maintain that a human organism that has lost a functioning brain, including the brainstem, is no longer an integrated unit of biological subsystems that constitutes a *living* organism – and for that reason cannot be identified with the original organism. The original body is dead.

Suppose that's right. Penelope's original body does not qualify as a living animal, despite the artificial maintenance of vital functions. Thus, if Penelope, the human animal, is still alive, she is for a brief time the extracted whole brain. Admittedly, the brain is a grotesquely mutilated animal, having lost most of its parts. But, one might argue, it is an animal nevertheless because it retains not only the capacity for consciousness via (primarily) the cerebrum, but also the capacity via the brainstem to power the vital functions of any body to which it is successfully attached. And the cerebrum and brainstem function in an integrated way. Following the brain transplant, the brainstem permits integrated functioning among itself, the heart, and the lungs, so Penelope has acquired a mostly new body. If this reasoning is sound, then the biological view accommodates the nearly irresistible intuition that one goes with one's mind in this whole-brain transplant case.[99]

Another possibility is that neither Penelope's original body, now brainless and maintained by mechanical assistance, nor Penelope's functioning brain is alive. Penelope, on this view, no longer exists. One could arrive at this conclusion by accepting that a brainless body cannot be a living organism while also holding that the functioning whole brain is likewise insufficient for continued life – basing the latter claim on the assertion that the brain isn't an integrated unit of biological subsystems. (In Chapter 4 I will argue that the brain does less, and a "brain-dead" body more, to integrate bodily functioning than is commonly thought, effectively undermining the present line of reasoning as well as that of the previous paragraph.)

---

[99] Van Inwagen reaches the same conclusion, although his reasoning, unlike mine, gives significant priority to the mental in determining what constitutes the surviving organism (*Material Beings*, pp. 173–4, 180–1).

Consider another way to view this case, corresponding to another way to understand human death. On this view, we should judge that Penelope's original body remains a living animal, even after losing its entire brain, because it retains the integrated functioning of subsystems that is characteristic of organisms. So Penelope is the artificially maintained, brainless body, and someone else gets Penelope's brain and psychological life.

What if, alternatively, Penelope's original body remains alive even though an isolated but functioning whole brain also constitutes a living animal? *Then we have a case of fission, Penelope dividing into two human animals – the brain and the brainless creature.* In that case, however much we may "identify" with the organ of consciousness, motivating us to say that the original human animal goes with her whole brain, the proper interpretation on the biological view is that the original animal, by dividing, has gone out of existence. But perhaps this is a way of going out of existence that is less bad than death, since this way permits *a kind of* survival – in two beings – and continuation of mental life in one of them.

In conclusion, given these various possible interpretations, it is not so clear (at least at first glance) what happens to the original human being in a whole-brain transplant case. The arguments of Chapter 4 will suggest that, on the biological view, one remains as the artificially maintained, brainless body. This, again, is counterintuitive. But even if my later analysis of the definition of death is incorrect, and the correct interpretation of the present case is that one would go as a mutilated organism with one's whole brain, the *cerebrum*-transplant case would remain as a theoretical thorn in the view's side.

2. *The Corpse Problem.* Several philosophers have challenged the biological view along the following lines.[100] Generally, the matter that constitutes your body at the end of your life will continue to exist and constitute a corpse for at least some time after you die. What we call a *corpse* after death seemingly existed before death as one's living body. According to the biological view, you are essentially a *living* human animal and therefore don't exist, following death, as a corpse. So what is your relationship to the future corpse?

---

[100] See, e.g., Baker, *Persons and Bodies*, pp. 206–8; and Carter, "Will I Be a Dead Person?" Carter rejects the thesis that we are essentially *living* animals, not the biological approach as a whole.

To see the problem clearly may require more precise terms, such as these provided by Olson:

1. When a human animal dies, there is ordinarily a concrete material object – its corpse – spatiotemporally continuous with it.
2. An animal's corpse doesn't come into existence when the animal dies; it exists and coincides with the animal before the animal dies.
3. An animal ceases to exist when it dies.
4. Hence, each human animal coincides, while it is alive, with a corpse-to-be that is numerically different from that animal.[101]

The argument's conclusion implies that you, an essentially living human animal, are numerically distinct from some other physical object associated with you right now: the corpse-to-be. Yet the biological view finds intolerable the thesis that you, who are essentially a person (according to Personalism), are numerically distinct from another being associated with you right now – the human animal. If the biological view must embrace two beings associated with you, it's hard to see why the implication that there are two beings associated with you should cast doubt on Personalism.

In response to this argument, we may distinguish three strategies. The first option is to deny premise 1 by denying that a corpse is a concrete material object. I find that utterly implausible.

A more congenial strategy is to deny premise 3: that an animal goes out of existence when it dies. Some who argue that we are essentially animals do not hold that we are essentially living and therefore maintain that we will exist as corpses.[102] To be sure, some of the things we say about corpses may seem to support this – think of the preacher in the *Kudzu* comic strip asking at a viewing, "Don't he look peaceful?", apparently referring to the man the mourners knew. But, when we're thinking carefully, I believe, we're strongly inclined to think that the corpse is not the individual we knew – who no longer exists, assuming there's no afterlife (or who exists elsewhere if there is). At least that's my intuition – when we die, we are no more – and it would take a lot, including the apparent absence of any other solution to the corpse problem, to convince me to abandon that intuition. Moreover, if we (and other life forms) survive death as

---

[101] "Thinking Animals and the Constitution View," Symposium of *Persons and Bodies. A Field Guide to Philosophy of Mind* (Spring 2001) [www.uniroma3.it/kant/field/bakersymp_olson.htm], pp. 3–4.

[102] See Carter, "Death and Bodily Transformation," *Mind* 93 (1984): 412–18 and "Will I Be a Dead Person?"

dead organisms – corpses – what exactly is an organism? I find it natural
to define *organism* in terms of certain forms of integrated functioning
constitutive of *life*, in which case *dead organism* is a logically misleading (if
sometimes convenient) locution.

The most credible solution, I think – namely, rejecting premise 2 –
focuses on the extraordinary differences that exist between a living animal and a corpse, despite a superficial similarity of outward appearances.
The fundamental differences concern what the living creature, but not
the corpse, *does*. The living organism resists the entropy characteristic
of inanimate matter by maintaining a system of integrated somatic functions, thereby maintaining a particular form or structure. Olson states
the idea nicely:

> A living thing, like a fountain, exists by constantly assimilating new matter, impos-
> ing its characteristic form on it, and expelling the remains. A corpse, like a marble
> statue, maintains its form merely by virtue of the inherent stability of its materials.
> The changes that take place when an organism dies are far more dramatic than
> anything that happens subsequently to its lifeless remains.[103]

On this view, although Gramma's corpse may look a lot like Gramma,
the two are numerically distinct and do not overlap in time, the corpse
coming into being at the point of death. This, I submit, solves the corpse
problem.

If by any chance this solution is incorrect, I think that is because (consistent with denying premise 3) we are essentially human animals but not
essentially living. If that thesis, which is an option within the broader biological approach, is correct, then there is no corpse problem: There is a
single human animal (and no other concurrent being or body), which is
first alive and later dead. So, if this turns out to be the preferable theory,
the biological approach will still be vindicated – just not along the lines
that I favor.[104] In conclusion, I suggest that the corpse problem is really an
intramural issue within the biological school rather than a point scored
for the psychological view.

3. Two Persons Associated with One Human Animal? Another challenge to
the biological view concerns cases where it appears that there are two (or
more) persons associated with a single human animal. If each of us is

---

[103] "Thinking Animals and the Constitution View," pp. 4–5.
[104] On the other hand, Chapter 4's analysis of human death depends on the thesis that we
are essentially *living* human animals.

essentially a human animal, then these cases feature just one of us, yet they sometimes seem to feature more than one of us.

Consider first Abigail and Brittany Hensel, conjoined twins whose two heads emerge from a shared torso, giving the irresistible impression that they are distinct girls.[105] This case does not seem to threaten the biological view, because it appears to involve the extremely rare circumstance of two overlapping organisms: They have two brains, including brainstems, two hearts, and two stomachs while sharing most of the organs located below the heart. The interpretation that the conjoined twins are two organisms fits comfortably with the intuition that the girls are distinct.

Posing more of a challenge to the biological view are cases involving conjoined twins in whom there is minimal duplication of organs yet clearly two centers of consciousness. Imagine a case in which there is one heart, one set of lungs, and so on, and perhaps even one brainstem, but the head is partly or entirely bifurcated due to the development of two separate cerebrums (each consisting of two hemispheres); crucially to our intuitions, two faces speak independently of each other. Here, it seems, there would be one human animal yet two persons and, more importantly, *two individuals of our kind*. If each center of consciousness, represented socially by one of the faces, is one of us, then since they cannot be most fundamentally human organisms – of which there is only one here – then neither can we. We, like they, would be essentially persons or subjects, according to the argument.

While there may be no actual cases like the one just described, there is another, well-documented type of case that arguably features more than one person associated with a single human organism: cases of *dissociative identity disorder*, formerly known as *multiple personality disorder* or, more popularly, *split personality*.[106] The interpretation of this phenomenon is enormously controversial.[107] But let us briefly consider a fairly typical case to get a feel for the relevant issues.

Imagine that James is the original person (and personality). At some point during or after a childhood damaged by repeated sexual abuse, another personality – who calls himself "Jimmy" – emerged and periodically

---

[105] For a discussion, see McMahan, *The Ethics of Killing*, pp. 35–9.

[106] American Psychiatric Association, *Desk Reference to the Diagnostic Criteria from DSM-IV* (Washington, DC: APA, 1994), pp. 230–1.

[107] See, e.g., Nicholas Humphrey and Daniel Dennett, "Speaking for Ourselves: An Assessment of Multiple Personality Disorder," *Raritan* 9 (1) (1989): 68–98; Stephen Braude, *First Person Plural* (Lanham, MD: Rowman & Littlefield, 1995); and Jennifer Radden, *Divided Minds and Successive Selves* (Cambridge, MA: MIT Press, 1996).

"took over" in guiding behavior. Insofar as James apparently cannot recall thoughts and actions that occurred under Jimmy's control, it seems to others that James is not conscious when Jimmy is. (In some cases of dissociative identity disorder, an alternate personality apparently has direct access to the consciousness of the "default" personality, although the latter lacks direct access to the alternate's consciousness. The challenge to the biological view is greatest when no personality has direct access to another personality's consciousness, so let us assume that this is so with James and Jimmy.) Much more cavalier, confident, and assertive than James, Jimmy typically emerges when James feels highly stressed or frightened.

Now suppose Jimmy promises to help you with apartment hunting next weekend. When the weekend comes, however, only James is available. He remembers making no such promise. Surely James is not bound by a promise Jimmy made, even if it was made with the utmost seriousness. This suggests, arguably, that Jimmy and James are different *persons* (not merely different personalities of a single person) despite their sharing a single human body. Moreover, it seems that there are *two of our kind* associated with one body – again suggesting that we are essentially not human organisms but persons or subjects, according to the argument.

In a case like this, in which mutual inaccessibility of consciousness between the alternates may be taken to indicate distinct centers of consciousness, it seems fairly reasonable to speak of two persons despite their association with a single body. The two-persons hypothesis receives further support from the appropriateness of (at least sometimes) treating the alternates in ways that are consistent with regarding them as distinct persons – for example, in not holding James to Jimmy's promise. Nevertheless, I do not think we should judge that this case features two individuals of our kind. Indeed, I doubt we should concede that there are two persons.

An alternative interpretation is that James is a mentally ill person whose consciousness is significantly compartmentalized. Rather than a person, "Jimmy" is a personality that represents a functionally enclosed mental "compartment" that James automatically (nonvoluntarily) switches to when he feels unable to cope with stress. The reason we shouldn't take Jimmy's promise to bind James is that James was incapable of autonomous decision making when the promise was made. Moreover, if psychiatric treatment eliminated the Jimmy personality or integrated it into James's self-concept, that would seem more aptly described as therapeutic to a single patient than as killing or otherwise destroying a real individual. In the end, it is probably more plausible to say that there is one person with a profoundly compartmentalized mental life rather than two persons.

Perhaps one could develop details of a case, not unrealistically, in such a way that makes the two-person hypothesis more compelling than I have suggested it is here. This hypothesis may be equally plausible in an extraordinary case of amnesia followed by the development of a personality that seems distinct from the original personality. (Interestingly, in these cases the new "person" always seems to retain certain skills, such as the ability to speak a certain language, that were acquired before amnesia – reminding us that some significant continuity of mental life ties the two personalities together. Similarly, alternate personalities in dissociative identity disorder do not have to learn language, basic reasoning, and general information about the world from scratch.) The biological view can, without incoherence, remain open to the two-person hypothesis. What would threaten the biological view are cases in which we seem required to say that there is more than one of our kind associated with a single human organism. I am not convinced that there are any actual cases of dissociative identity disorder, amnesia followed by a new personality, or conjoined twins whose plausible interpretation requires us to conclude that there are two of us.[108]

More challenging to the biological view are certain cases that, although hypothetical, seem no more farfetched than cerebrum transplants. The most challenging, I think, is the case of conjoined twins where there is clearly only one human animal whose two faces speak independently of each other. If you said, "We are going to amputate you [tapping one face] to improve the health of the overall animal," it seems unlikely that the individual represented by that face would identify with the animal and feel relieved at the prospect of therapeutic surgery. Rather, she would presumably fear her destruction. And this may seem to suggest that there are not only two persons, but two individuals of our kind, associated with that human body.

But this inference would be hasty. Another plausible interpretation is that in this case, what matters prudentially diverges somewhat from

---

[108] I would say the same for cases involving split-brain patients, who appear to have two separate centers of consciousness. For an engaging introduction to this phenomenon and the issues it raises, see Thomas Nagel, *Mortal Questions* (Cambridge: Cambridge University Press, 1979), ch. 11.

numerical identity. In this freakish case, there is only one of us, one human organism, but she has two centers of consciousness, each with its own interests. Such a bifurcating of consciousness and welfare is strange, to be sure, because it very rarely if ever happens in the real world, but such a phenomenon does not seem impossible.

4. An Allegedly Poor Fit with the Practical Concerns Associated with Identity. A final major challenge to the biological view is the charge that it is a purely metaphysical view that fits poorly with our practical concerns. This charge is most likely to be advanced by someone who believes that the psychological view captures our practical interests especially well and who plays out farfetched hypothetical cases, like that of the prince and the cobbler, in standard Lockean ways. As we have seen, though, the psychological view has major difficulties that suggest its overall inadequacy. Moreover, even if that approach accounts plausibly for certain practical concerns, it doesn't follow that the biological view does not. Third, as we have seen, the practical relevance of farfetched hypothetical cases is pretty dubious. But let us address the present challenge to the biological view with greater specificity by identifying particular charges and the replies the biological theorist can give. In doing so, I will construe the idea of practical concerns broadly and include (1) self-knowledge of identity over time, (2) belief in an afterlife or in reincarnation, and (3) the thesis that identity is not what matters in survival. (In an earlier section we treated these separately from what were called practical concerns.)

Charge 1: People have self-knowledge over time. But the resources that permit one to identify oneself as the same person over time – memory and anticipation – require psychological continuity.

Reply 1: That's true. But since, in the world as we know it, psychological continuity requires the continuation of a particular biological life, self-knowledge requires the latter as well. To be sure, continuation of a person's life isn't sufficient for self-knowledge – which a PVS patient clearly lacks. But so what? Let's not confuse personal identity with knowledge of it.

Charge 2: Moral agency and therefore moral responsibility presuppose psychological continuity.

Reply 2: But, again, in the world as we know it, psychological continuity requires continuation of the agent's (biological) life. Nor does it matter much that one may continue to live without any longer being morally responsible – say, if one is severely demented or comatose. After all, early in life, one achieves psychological continuity over time, at some point in infancy, before having any moral responsibilities. There is no reason to think that a theory of human identity must fit *perfectly* with moral agency. And the present view fits reasonably well with the latter.

Charge 3: Acting intentionally, making life plans, and watching after future interests presuppose that one persists through time and will be around to carry out plans and benefit from prudence. More specifically, these fundamental aspects of human life assume that each of us will be *the same person* at the later times in question – vindicating the psychological approach.

Reply 3: Once again, in all actual cases with which we are familiar, to be the same person requires continuing a single biological life. Of course, part of the charge is that continuing a life, though perhaps necessary, is not sufficient for capturing our interests in planning and prudence, psychological continuity also being necessary. But I contend that planning and prudence only *paradigmatically* presuppose sameness of person over time rather than having the latter as a strictly necessary condition. Just as I can imagine someday being so demented as to be psychologically discontinuous with myself now (being, that is, a sentient nonperson), I have concern that *I* be treated well should I fall into this state. So, at most, planning and prudence require sameness of subject (sentient being), not sameness of person. And even this more modest requirement is not self-evident; many thoughtful people believe they would have interests in a state of permanent unconsciousness prior to death, such as interests to avoid great indignity and to have one's previously expressed wishes honored. Finally, even if such people are wrong, and our interests in planning and prudence strictly require the persistence of a subject, in the real world there is a very strong correlation between remaining alive and maintaining subjectivity. So the biological view does reasonably well on this score, too.

Charge 4: Social life among human beings involves treating each other as persons with ongoing narratives and therefore psychological continuity.

Reply 4: That's roughly correct. Then again, we identify and recognize others largely by recognizing their living bodies, even if we almost always assume that a particular body embodies a continuing person. Moreover, this last assumption is not universal, since we also socialize with prepersonal newborns and some demented nonpersons.

Charge 5: Only the psychological view fully captures interest in an afterlife and somewhat less common interest in a beforelife, inasmuch as psychological continuity between our present life and a time before birth or after death would convince us that our present life wasn't the whole of our existence. The biological view, by construing human identity as sameness of biological life, precludes reincarnation and an afterlife.

Reply 5: Yes, as discussed earlier, the biological view precludes these phenomena, for one's biological life and therefore one's existence begins at around the time of conception and ends at death. Even if the existence of our mental life before (the most recent) birth or after death is, or seems to be, logically possible, there is ample reason – given the reasonableness of naturalism – to think it metaphysically impossible. Consider stories of an afterlife that strike us as compelling. What makes us think that an afterlife is logically possible and that psychological continuity following death would convince us of persisting is the first-person perspective built into the thought experiments. We are asked, "What would *you* do if you found *yourself* awake, in a wonderous unearthly place, after *you* died?" But this phrasing begs the question of identity. A better phrasing is, "What should we say about a situation in which someone were conscious, existing in a state fundamentally different from that of life on earth, and believed herself to be continuing the psychological life of someone who recently died?" Well, if such a state of existence were possible (and, again, from a naturalistic perspective it seems impossible), perhaps this individual would be systematically deluded, like Parfit's replica on Mars or a madwoman who thinks she's Hilary Clinton. And even if such a state were possible, and such psychological continuity – with the right kind of cause and no branching – were sufficient for persistence, it's doubtful that such continuity would be necessary; persistence of a bare subject of consciousness should also suffice. (And what would the bare subject be? Something possessed before death, since it has to be the same bare subject: not the brain, which death often destroys, so probably a soul. But if the view reduces identity to sameness of soul, it abandons the Lockean program.)

Charge 6: Identity itself is not what primarily matters in survival. What most matters, prudentially, in our continuing existence is psychological continuity – which, conceivably, can exist without identity. Consider: Although fission presumably ends the existence and identity of the original person, the prospect of her fissioning should not be considered nearly as bad as ordinary death, because psychological continuity is maintained (in two persons).[109] So the biological view offers a criterion of human

---

[109] Martin presents a scenario he calls *fission rejuvenation*, arguing that it demonstrates more convincingly – than does fission as ordinarily characterized – that identity is not what

identity that doesn't capture what most matters in survival. Now, the possibility of a divergence between identity and what matters prudentially is most evident in the case of a (whole) cerebrum transplant that succeeds in maintaining psychological continuity. Not only are we strongly inclined to think that one goes with one's mind into the new body; we strongly tend to believe that what most matters prudentially is psychological continuity, or at least the capacity for consciousness, and that the cerebrum takes this into the new body. But the biological view, by implying that we do *not* go with the cerebrum, separates identity from what most matters in survival. By construing identity in a way that has nothing to do with psychological continuity, the biological view proves to be a purely metaphysical thesis with no necessary connection to the sorts of practical concerns that interest us in identity in the first place. (This last assertion broadens the concern about what matters to embrace all the practical concerns discussed in this section.)

Reply 6: The biological view is a theory of human identity, of our persistence conditions. As such, it is a metaphysical and conceptual theory. Strictly speaking, then, it is not responsible for tracking all of the concerns we tend to associate with identity. Moreover, the present charge greatly exaggerates the divergence of the biological view from what matters. *In the real world as we have experienced it so far, moral responsibility, planning and prudence, and the other practical concerns under discussion all presuppose or depend on our continued existence; so does what matters in survival, even if we agree that what most matters in survival is psychological continuity or the continuing capacity for consciousness.* So, in cases known to be actual, the biological view does not diverge from what matters.

Cerebrum transplants might be possible someday. (Whole-brain transplants may also prove possible, but we will not consider them here because the biological view's implications in this sort of case are more debatable.) Quite arguably, in a cerebrum-transplant case, what most matters to us would diverge from identity.[110] For example, one might well prefer, without any clear lapse in rationality, (1) having one's cerebrum transplanted into a new body as one's original body (and, on the biological

primarily matters in survival (*Self-Concern*, pp. 54–7). I cannot take up his arguments here.

[110] Elsewhere ("Identity, Killing, and the Boundaries of Our Existence," *Philosophy and Public Affairs* 31 [2003], pp. 424–5) I argued that, our intuitions notwithstanding, because you would not be the recipient of your cerebrum in the transplant case, you would lack strictly egoistic reasons for caring about the recipient's welfare. Now I am unsure about that thesis.

view, oneself) is permitted to die to (2) surviving with a quality of life considerably lower than that promised in the transplant scenario, even if not so low as to make life not worth living. Now, if fission (as the psychological view understands it) becomes possible, that is likely to involve transplanting brain hemispheres into two brainless bodies; copying precise mental contents and downloading them into blank-slate brains, for example, is far less likely to be achievable. But if transplanting a whole cerebrum is sufficient to separate identity from what matters, fission via hemisphere transplant adds nothing important to the issue. Indeed, it may be less convincing as an example, because a person with only one hemisphere is likely to have much less mental functioning than someone who has both hemispheres.

As discussed in the previous subsection, another type of case may separate identity and what prudentially matters. In a hypothetical case involving conjoined twins in which there is clearly just one human organism but two centers of consciousness, there would be (according to the biological view) only one individual of our kind yet two socially distinguishable faces governed by distinct minds with distinct interests. In short, there would be one of us but two centers of prudential concern.

So what we find, ultimately, is a small possible divergence between identity and what matters – and, then, only if we consider presently hypothetical cases. I suggest that this is an acceptable price to pay for a theory that otherwise seems more coherent, more metaphysically plausible, and more consonant with educated common sense than its competitors.

5. *Conclusion.* On the whole, the biological view can respond fairly well to the major challenges that confront it. The cerebrum-transplant intuition and the hypothetical conjoined-twins case remain standing as challenges. Other things equal, a theory is stronger if it can avoid contradicting a persistent intuition such as the judgment that you would go with your functioning cerebrum, and if it can avoid the oddity of asserting that one of us could have two centers of prudential concern. Meanwhile, the corpse problem has been satisfactorily addressed. If the chief claim in my rebuttal – that the living human body and the corpse are numerically distinct entities – is incorrect, then that simply vindicates a different variant of the biological approach: the view that we are essentially human animals (first alive and then dead). Finally, we found that the biological view's understanding of human identity does not diverge radically from what matters to us in survival or from our practical concerns more generally. Interestingly, where a divergence seems most likely is in the case of a

cerebrum transplant. (The hypothetical conjoined-twins case seems less compelling because it is harder to interpret.) It should not be surprising, therefore, that one theorist has embraced a theory of our identity that takes the biological view as a point of departure but modifies it to account for intuitions about whole-brain transplants – and, by implication (considering how he interprets whole-brain transplant cases), cerebrum transplants.

A SEMIBIOLOGICAL ALTERNATIVE: HUMAN BEINGS AS A PARTLY PSYCHOLOGICAL KIND

Offering a view of our identity that differs both from the psychological view and from a strictly biological view, Mark Johnston asserts that we are *human beings* in the following technical sense:

> [We are] beings which necessarily are normally constituted by human organisms, and whose conditions of survival deviate from those of their constituting organisms only because a human being will continue on if his mind continues on, so that a human being could be reduced to the condition of a mere brain so long as that brain continues the human being's mental life. Although having all but its brain destroyed is too much for a human organism to survive, a human being might plausibly be held to survive this.[111]

To understand the basis for this view, let us consider Johnston's methodology.

From the naturalistic framework, we begin with the assumption that we are human organisms. This assumption is not only supported by science and common sense; it also permits us to vindicate our ordinary, highly reliable practices of reidentifying other human individuals, which we do by recognizing particular human bodies. When a person reidentifies herself, however, she relies on memory. One remembers doing something or having an experience; and, by its nature, memory comes with the belief that the earlier subject is the present subject.[112] But how could a mental faculty like memory guarantee that some biological criterion of identity is fulfilled? Memory links current mental states to past mental states, in nondelusional cases ensuring sameness of *mind* – which, again, we take to be a brain functioning in certain ways. But now consider the transplant intuition: that one goes where one's mind, or functioning brain, goes. The person who wakes up postoperatively with my brain seems to remember

---

[111] "Human Beings," p. 64.
[112] Ibid., p. 77.

my life before the transplant. Can we, in good theoretical conscience, accommodate his belief that he is I and the transplant intuition more generally?

Johnston thinks so. For "if anything deserves the name of a conceptual truth about the relation between persons and minds, it is the claim that a person cannot be outlived by (what once was) his own mind."[113] This has a direct implication for brain transplants: Since your mind is your functioning brain, when the latter is transplanted, so is your mind, from which it follows that you go where your brain goes. Johnston is careful to point out, however, that this modification of the biological view does not imply that one could survive teletransportation or the copying and downloading of one's mental contents as one's original body is destroyed. As argued earlier, a mind is not an immaterial thing – or a set of mental contents – that can jump from body to body, so in these cases it is *illusory* that someone's mind has traveled. As Johnston sees it, the view he proposes is close to the biological view that we are essential human organisms: "So far in every actual case a human being is constituted by a particular human organism, and so the survival of the organism is for all practical purposes a necessary condition of the survival of the human being."[114]

This modified biological view enjoys the advantages of a naturalistic framework. It also accommodates the transplant intuition that we go where our functioning brains go and, for precisely this reason, its criterion of identity fits even better than that of the (pure) biological view with what seemingly matters in survival. Overall, Johnston's view may seem highly attractive.

Nevertheless, it has its share of difficulties. Before addressing them, however, we should note some unclarity about what the view comes to. Johnston states that survival of the organism is, in actual cases thus far, necessary for survival of the human being. Is survival of the organism also sufficient for survival of the human being? If not, then this view begins to resemble Baker's in critical respects – for example, in holding that a human being would not survive in PVS. Is it Johnston's view that we come into being as the human organism does, which is to say, as fetuses? If not, then presumably we come into being as minded beings so that human beings are, like persons on Baker's view, essentially minded. Then the view would inherit from Baker the fetus problem (though that is arguably not

---

[113] Ibid.
[114] Ibid., p. 79.

a serious difficulty) and, unless bare sentience suffices for being a human being, the newborn problem. It would also imply that, strictly speaking, we are not animals, being merely constituted by them. On the other hand, it would avoid some of the difficulties that stem from Baker's reification of the mind – or first-person perspective – by identifying the mind with the functioning brain, setting an important reality constraint on the thesis that we go where our minds go.

Another reading of this hybrid view is that each of us is simply a human organism until she acquires a mind, at which point survival of the mind suffices, but isn't necessary, for her survival. Upon gaining a mind, one can continue either by the human organism's continuing or, in the case of a brain transplant, by the mind's continuing. But this view might be incoherent. Suppose both the organism and the mind survive but part ways: Only the cerebrum is transplanted and the original organism survives, with the brainstem powering vital functions. Each continuer satisfies what we've supposed is a sufficient condition for surviving. Could we count this as a bizarre case of fission? Not if we take Johnston at his word: "a person cannot be outlived by (what once was) his own mind." This implies that one would go with one's functioning cerebrum. And that seems to fit with the interpretation that brings Johnston's view pretty close to Baker's: We are essentially minded things and therefore do not exist when a human organism lacks a mind.

An additional charge against this view is that it seems ad hoc. Johnston begins with the promising biological view and then, when it faces the challenge of brain transplants, introduces a modification to accommodate intuitions about our identity and what matters in survival. Further, the departure from the biological view resulting from the theoretical jerry-rigging puts the scientific respectability of the theory in doubt. *Human being*, in Johnston's sense, is a much less plausible candidate for a natural kind – a kind determined by nature[115] – than is *human animal*, defined purely biologically; such concepts as *Homo sapiens, hominid,* and *primate* are employed in every biology textbook, whereas concepts like Johnston's *human being* are unheard of. Then again, Johnston indicates that he is investigating *our conception of what we are*, and is not necessarily looking for a "substance kind" that determines our essence.[116]

---

[115] For classic discussions, see Saul Kripke, *Naming and Necessity* (Cambridge, MA: Harvard University Press, 1970); and Hilary Putnam, "Meaning and Reference," *Journal of Philosophy* 70 (1973): 699–711.

[116] "Human Beings," pp. 64–5, note 8.

Fair enough, but, as Johnston's own comments suggest, identifying our conception of what we are is not simply a matter of tracing out our intuitions; it involves disciplining those intuitions by considering the reliability of their source and requiring them to conform to our knowledge about the real world. Now the impetus for Johnston's departure from the original biological view is the transplant intuition plus his understanding of our practice of reidentifying ourselves over time. The transplant intuition announces that we go with our minds, while consideration of self-reidentification announces that we remember our own past. But let us take seriously one of Johnston's most valuable cautions about the intuitive case method, which can extend both to the transplant case and to our practice of self-reidentification: *Be careful not to overgeneralize from ordinary cases to every possible case.* In ordinary cases, one goes where one's mind goes and one remembers only one's own past. But cerebrum transplants, in our world, count as extraordinary cases that haven't even occurred yet. Perhaps, then, if there are good theoretical grounds – such as all the arguments for the biological view and against the psychological view – for saying that we are essentially human organisms, we should interpret the cerebrum-transplant intuition as an overgeneralization from ordinary cases (in which we go where our minds go) and the resulting individual's apparent memories of preoperative life as systematically deluded.

In conclusion, Johnston's hybrid view clearly has major difficulties as well as advantages. I suggest that it is less coherent and compelling than the original biological view. If so, then Johnston's methodological suggestions prove more valuable than his view of what we are.

ONE MORE ALTERNATIVE: MIND ESSENTIALISM

Johnston claimed that one necessarily goes where one's mind goes. This thesis underlies one more theory of our identity and essence that merits our attention. This theory differs from Johnston's in offering a distinct account of the relationship between the human animal and person, in presenting a single criterion of human identity (Johnston half-suggests two criteria), and in being more theoretically explicit.

In what he calls his *embodied mind account,* Jeff McMahan argues that we are essentially minded beings or minds.[117] But minds, he argues,

---

[117] *The Ethics of Killing,* pp. 66–9. Parts of the following description and critique of this account draw significantly from my "Identity, Killing, and the Boundaries of Our

are – or are caused by – brains (more precisely, those brain parts in which consciousness is realized) functioning in certain ways. And we can plausibly individuate minds not in terms of their mental contents, but by individuating brains. Thus his *mind essentialism* suggests that we are essentially *embodied* minds. Accordingly, the "criterion of personal identity is the continued existence and functioning, in nonbranching form, of enough of the same brain to be capable of generating consciousness or mental activity."[118]

Like Johnston and me, McMahan holds that the intuitive case method largely supports the thesis that we are essentially bare subjects – to use his terms, minds – not the thesis that we are essentially persons, or beings with continuity of experiential contents over time. Thus, in his view, which depends heavily on the intuitive case method, continuation of the *capacity* for consciousness (which requires a persisting, functioning brain) suffices for our continued existence. In this way, he parts ways with Parfit and many other psychological-continuity theorists.

One major advantage of the embodied mind account is that it accommodates the intuitions that one would go with one's transplanted cerebrum and that what would most matter to us, prospectively, is what would happen to the individual who acquired the cerebrum. By understanding the mind in terms of the brain – at least on the specification of the view in which the mind *is* the functioning brain – it also apparently avoids both the reification of mind of which Baker's view is guilty (though I will revisit this issue momentarily) and such mysterious criteria of identity as sameness of first-person perspective. Moreover, since newborns have the capacity for consciousness, it avoids the newborn problem: We came into existence when the fetus acquired the capacity for consciousness or sentience, perhaps at around five months, and therefore existed at birth. On the whole, this account may represent the most promising version of the psychological approach. (Asserting that we are essentially minds

---

Existence," pp. 419–21. Peter Unger also offers an embodied mind account in *Identity, Consciousness, and Value.* For Unger, you persist if your core psychology – your capacities to experience consciously, to form simple intentions, and to reason in rudimentary ways – is continuously encoded in a physically continuous realizer (ordinarily, your brain) or a succession of physical realizers. This theory's advantages include avoiding the reification of minds and not requiring for identity the continuity of mental contents. But I find McMahan's version of the embodied mind approach more powerful, in part because the bare capacity for consciousness seems, for reasons discussed earlier, to be the most plausible *psychological* criterion of our identity. Of course, I think a biological criterion is preferable.

[118] *The Ethics of Killing,* p. 68.

or bare subjects, not that we are essentially persons, the embodied mind account is not quite a variant of Personalism.)

But can this theory respond adequately to other challenges we have considered? First, regarding the fetus problem, mind essentialism implies that we were never *presentient* fetuses, but that implication is arguably not damaging. While I find it more plausible to say that we were such fetuses, many people disagree. Next, the theory faces the challenge of explaining the relationship between you, the person, and the presentient prede-cessor, as well as your relationship to the (possible) future nonsentient human animal in PVS. Interestingly, McMahan rejects the constitution view of these relationships in favor of a part-to-whole thesis that explains our relationship to "our" organisms at any time. He contends that you, the person or embodied mind (roughly, the brain), are *part of* the human animal or organism with which you are associated.[119]

Doesn't this imply that there are now two conscious beings reading these words – the person and the human animal? As discussed earlier, that would seem one too many. The biological view can claim, more com-fortably, that there is exactly one conscious being – the human animal – which, at the present time, has properties constitutive of personhood and is reading. But McMahan contends that the embodied mind account's implication that there are two conscious being is not so strange in view of certain analogies. When you honk your car's horn, it is correct to say both that the horn makes a noise and that the car makes a noise (the same noise). When a tree's limb grows, the tree also grows. A complex thing may do *X* in virtue of one or more of its parts' doing *X*. In this way, we may say that the human animal is conscious and so are you, the per-son, without great oddity: The organism is conscious in virtue of having a mind that is conscious.[120]

Thus we may say the following about your relationship to the presen-tient fetus, the organism in PVS, and human animals more generally. The human organism, which began as a presentient fetus, acquired the capacity for consciousness (or sentience). When it did so, you – the em-bodied mind – came into being as part of the organism, specifically (and roughly) its brain. The organism grew and developed, as did its part that is you. In the future, the organism may lose its capacity for consciousness – at which point you, the person, will die – before it, the organism, dies. In general, although you are not an animal, you are part of an animal.

---

[119] Ibid., pp. 89–92.
[120] Ibid., pp. 92–4.

In this way, you and your animal are intimately related but distinct, with different criteria of identity.

This account of our identity and essence has enough merits to count as a contender. I nevertheless maintain that it is less plausible and less theoretically well motivated, on the whole, than the biological view. My critique of the embodied mind account focuses on some of its implications, its inability to explain cogently what we are, and considerations of methodology.

First, consider its implications. The embodied mind account implies, once again, that none of us is an animal. I cannot believe this. I do not think that biology teachers systematically misinform students when they teach them that each of us is an animal. (Not that biology teachers are experts in metaphysics; rather, I assume that the natural sciences play a leading role in informing us about ontology.[121]) If scientifically informed common sense is right, then we are animals – not beings constituted by animals or parts of animals. McMahan allows that *animal* is a substance concept, so he allows that there are two conscious beings thinking my present thought: the animal and the person. Now I still find the two-beings implication somewhat counterintuitive, although perhaps tolerable given his part-to-whole analysis. The problem, to my mind, is thinking that *person* – or, more precisely, *mind* – is also a substance concept.

Again, for all we know, supercomputers may someday achieve the capacity for consciousness. And there are probably conscious extraterrestrials somewhere in the universe. Yet human beings, conscious extraterrestrials, and supercomputers are, or would be, so materially and structurally different from each other that it seems more plausible to regard *mind* not as a substance concept but as a concept that identifies a common function – the capacity for consciousness – among what may turn out to be different basic kinds. This point is even more evident if we include, among the set of logically possible minded beings, factually impossible beings such as angels, devils, and immaterial spirits.

Further, if we are not animals, then what exactly are we? I don't think McMahan, who professes neutrality on the mind–brain relation,[122] can produce a tenable answer. Consider the possibilities.

McMahan can't claim that the mind is an irreducibly mental (immaterial) substance, entailing the thesis of substance dualism, because he

---

[121] Here I make common cause with John Searle (*The Construction of Social Reality* [New York: Free Press, 1995], pp. 5–7).

[122] *The Ethics of Killing*, p. 88.

attempts to refute that thesis.[123] Nor can he plausibly assert *property* dualism, which states that while all substances are material, some properties – the conscious ones – are not. Conceptually, the mind is *that which is* conscious, the subject of consciousness. If this concept of mind is, in fact, realized in a distinct entity (as McMahan holds and I, claiming reification, deny), then the mind is an entity – a substance – that has properties; it can't itself be reduced to a set of properties. If the mind could be so reduced, then on McMahan's view you are not a substance at all but merely a set of properties. That would contradict his explicit assumption that you, the mind or person, are a substance – and a stranger thesis is hard to imagine! So the mind, on his view, must be something physical.

McMahan might claim that the embodied mind is the brain (more precisely, certain brain parts). But then his mind essentialism will imply, implausibly, that you weigh less than ten pounds, you can't walk (though you can make your organism walk), you are gray and convoluted earlier than you expected to be, and so on. Moreover, brains can exist in a non-functional state in corpses, no less than fingers can. So, if we are brains, we could exist without the capacity for consciousness – contradicting McMahan's identity theory.

Perhaps, instead, we are brains that have the capacity to generate consciousness: *functional* brains. But assuming we are substances and the brain is a substance, this thesis would imply that my brain and my functional brain are *two distinct substances* – and that is hardly plausible.[124] So it seems that McMahan lacks a cogent account of what a mind is, and therefore of the kind of thing we are.[125]

Methodologically, I suggest that the embodied mind account relies somewhat too heavily on the intuitive case method and is not fully satisfying from a naturalistic standpoint. To its credit, if interpreted as I suggest, it is naturalistic enough to understand our mental lives in terms of our brains, avoiding the reification that plagues many psychological theories. But, in my estimation, it doesn't take seriously enough the naturalistic

---

[123] And quite persuasively (ibid., pp. 14–24).

[124] Eric Olson has presented roughly the same argument orally. Assuming you are a substance, one might wonder why the admission that the brain is a distinct substance doesn't create a two-substances problem similar to that attributed to the constitution view. The reason is that what was problematic was the idea that there are two conscious beings or substances associated with exactly the same matter, our entire bodies.

[125] Again, I regard it as reifying to refer to minds at all as if they were distinct things. Talking about minds is a convenient way to talk about brains while they are functioning in familiar ways. Since obviously *we* (whatever we are) are substances, mind essentialism cannot be reconciled with this point about reification.

judgment that we are animals. Meanwhile, as discussed earlier, the intuitive case method runs the risk of overgeneralizing from ordinary cases,
in which continuation of a mental life gives us sufficient reason to assert
that someone has persisted, to extraordinary cases that have yet to be
encountered.

I suggest that, in cases in which we intuitively believe that one would
part ways with the original human animal, the embodied mind account
conflates our *numerical identity* and our *patterns of identification*. Yes, we
would tend to identify with the transplanted cerebrum or the individual who acquires that cerebrum, but that doesn't show that we would be
identical with it or her. Because we identify so strongly with our mental
lives, considering them so important, we identify with – and would even
like to be – whoever has (what was originally) our cerebrum. But let's not
conflate metaphysics and value. In Chapter 3, we will discuss a sense of
identity that presupposes numerical identity – namely, *narrative* identity –
for which our continued capacity for consciousness proves crucial in ordinary cases. (And in Chapter 5 we will return to the topic of identification.)
But, if the arguments of this chapter are correct, the continued capacity
for consciousness is not necessary for our numerical identity.

CONCLUSION

The reflections of this chapter have two major upshots. First, we have
rejected Personalism, according to which we are essentially persons and
our numerical identity consists in some form of psychological continuity
over time; we have similarly rejected the more modest thesis that we are
essentially beings with the capacity for consciousness. And, more generally, we have found the psychological approach unpromising. Second, we
have defended the biological view, which holds that we are essentially human animals and that human identity consists in sameness of biological
life.

One way to grasp what a view comes to is to note its judgments about
particular cases. While we have argued in this chapter that the intuitive
case method is somewhat unreliable and should not be the driving force
behind a theory of what we are, we have ventured judgments regarding
the biological view's implications for identity in a range of cases, both
real and hypothetical. Let us imagine, for each case, the question "Would
you survive?" or, in the fetal case, "Did you exist then?" (For the cases of
conjoined twins and dissociative identity disorder, which do not address
these questions, see the subsection "Two Persons Associated with One

Human Animal?") The following summary of judgments and supporting rationales may helpfully amplify our characterization of the biological view.

- *Presentient fetus*: Yes, you, who are a human animal, were once such a creature.
- *Severely demented, sentient nonperson*: Yes. A human animal can exist without being a person.
- PVS patient: Yes, you would survive, because the human animal continues to live.
- *Teletransportation in which the original body, including the brain, is destroyed*: No. The human animal is killed. We can't say his mind went into the new body on Mars, because a mind is not a spiritual substance or an abstract set of mental contents; to speak of a mind is to speak, in a somewhat reifying way, of a brain functioning in certain ways. This case, by the way, may be forever technically impossible. However, as long as it's not described in a way that entails that one is indeed teletransported to Mars, or that one's mind travels to Mars, it is logically possible.
- *Kafka-like transformation – say, into a fly – with apparent psychological continuity*: This scenario is physically or metaphysically impossible. Flies cannot be mentally continuous with previously existing humans because flies' brains are not complex enough to support such thought. Nor could a human body transform into a fly body. If one insists that we try to imagine all this happening anyway, one might as well be open to a human transforming into a chair that seems to remember the human's life. If forced at gunpoint to address this case, I'd say that nothing that is originally a human animal could ever take the form of a fly or chair.
- *The prince and the cobbler (where you begin as the prince)*: Yes, you'd survive – in your original body. You would delusionally think that you are now the cobbler. As with teletransportation, one's mind doesn't leave one's original body. After all, one's original brain is still functioning enough to support consciousness, just not very well! But, even if your brain stopped supporting consciousness, you'd survive as the prince so long as the prince-body human animal lived. As Locke describes the case, one would like to know what the mechanism of psychological changes in each body is supposed to be, bringing us to the next case.
- *Apparent body switching (à la Williams) in which A's and B's mental contents are recorded and downloaded into the B body and A body, respectively*: This

is another one that may forever be technically impossible. But it is logically possible. The verdict here is the same as for the immediately preceding case.

- *Gradual replacement of all of one's organic body parts with inorganic parts, maintaining psychological continuity (à la Baker)*: We encountered this case but didn't offer a verdict. Since we do not know whether it's possible for inorganic material to support consciousness, we don't know whether this case is metaphysically possible. But it's logically possible. Since you are essentially a human animal, the question is whether the resulting person is the same human animal and continues your biological life. Certainly you would survive replacement of your amputated leg with a prosthetic leg and replacement of your diseased heart with an artificial heart. But wholesale replacement of your body parts with inorganic parts may be different. If an animal is necessarily (largely) organic, then you do not survive the changes. (Since it's doubtful that there's a precise point at which you'd cease to exist, this judgment would support the thesis that identity is indeterminate.) If, alternatively, an animal is anything that begins as an animal and continues to live, where living is interpreted in a functional (rather than an organic) way, then you would survive the changes. I'm somewhat inclined to judge that an animal is largely organic, meaning you don't survive according to the biological view. That you wouldn't survive this, given the psychological continuity between you and Tin Woodsman, strikes me as at least slightly counterintuitive – but less so than the biological view's verdict in the cerebrum-transplant case. I don't think this case poses a significant problem for the biological view for three reasons: (1) uncertainty about whether inorganic material could support consciousness; (2) uncertainty that an animal could not become entirely inorganic; and (3) unclear intuitions about whether you would survive such changes.
- *Whole-brain transplantation in which the original body continues to function with mechanical assistance*: It's not crystal clear what the biological view implies here about your fate. If it's correct to say that an entire brain, including the brainstem, can be a (radically mutilated) organism if separated from the rest of the body – which would be dead – then the biological view implies that you go with your whole brain into a different body. If, as I believe, that's incorrect, then the biological view implies either (a) that you remain with your original, now brainless, living body, (b) that you die, or (c) that this is a case of fission, on the supposition that cardiovascular functioning and whole-brain

functioning both suffice for survival. On the strength of arguments presented in Chapter 4, I support option (a).

- *Cerebrum-transplant case, in which the original body retains the brainstem and continues to live.* You survive as the original body, which now lacks its organ of thought. The person who now has your cerebrum falsely believes that she is you.

This chapter has developed a view about the numerical identity and essence of human persons. But it has not discussed identity in another major sense of the term, *narrative identity.* It has also not investigated related issues concerning what we can make of ourselves, or *self-creation.* These issues, which are crucial to understanding *who you are* – in one familiar sense of this phrase – comprise the subject matter of Chapter 3.

One final thought before turning to Chapter 3. The issues we have tackled in this chapter are very controversial, and I can hardly hope to persuade every reader that the biological view is correct. Importantly, although I will assume the correctness of this view in the remaining chapters, *the main arguments of Chapter 3 and many specific arguments in later chapters will not depend on its correctness.* Thus, for example, most of what I argue regarding advance directives in Chapter 5 depends on my claim that person essentialism is false but not on my claim that mind essentialism is false. Most of what I argue regarding enhancement technologies in Chapter 6 depends on clearly distinguishing numerical and narrative identity (as most theorists have not done), but not on the biological view in particular. And, in Chapter 7, whereas my arguments concerning our origins, prenatal identity, and the ethics of prenatal genetic interventions rest squarely on the biological view, the discussion of the nonidentity problem is relatively independent of this view. Moreover, in the discussion of abortion, the chief role of the biological view is to set up a formidable challenge to the liberal position I defend; and much of the argumentation rests on an account of prudential value, the Time-Relative Interest Account (defended in Chapters 5 and 7), which is logically independent of the biological view. On the whole, the positions developed in Chapters 4 to 7 depend on the distinction between numerical and narrative identity, the biological view of numerical identity, Chapter 3's framework for understanding narrative identity, and/or the Time-Relative Interests Account – but rarely on all of these.

3

Human Persons

Narrative Identity and Self-Creation

Erik wasn't sleeping well. In fact, several hours before he went to bed each night, a feeling of dread about going to work the next day began to grow in him. He hated his job as a paralegal: the notion that time was well spent only if it could be billed to a client; the harassing (and lying) letters to members of the sheet-metal trust fund saying the law firm would sue in ten days if debts were not paid to the fund; phony lawyers pretending to care less about money and more about humanistic concerns than they did; the rigid hierarchy; the way lawyers procrastinated and then had to pull consecutive all-nighters right before a brief was due, spreading misery to everyone who worked with them; the use of the cliché "This is the real world" to justify every unethical act; the cruelty toward vulnerable coworkers such as secretaries and newcomers like Erik. No, Erik wasn't sleeping well. He was moderately depressed and very anxious. And his mind worked overtime when he was supposed to be sleeping.

"Should I quit my job?" he asked himself. "But I've long planned to go to law school. Assuming I do, wouldn't it make sense to get more legal work experience? And, if I can't handle my first job, what are my prospects for a legal career? Then again, maybe I'm just not cut out for work in corporate law. There are lots of things you can do with a law degree, like working for the government or a nonprofit firm; I've heard public defender positions can be satisfying and less intense than jobs in private firms. I'm obviously a good fit for law school, being strong in the humanities – yet also practical.

"Or is it obvious? Maybe I've always assumed that because Dad and Mom both went to law school. I majored in history, and I really love

77

medieval history. But you can't pursue that! Not without a lot of risk, anyway. Think of it: five or more years of grad school with a very real chance of unemployment or only part-time work afterward. No wonder Dad discouraged me from considering grad school! But... there I go again: taking my parents' point of view. What do *I* want? What would make *me* happy and fulfilled? Who the hell am I, anyway?"

Unlike Penelope from Chapter 2, who was preparing to undergo a cerebrum transplant, Erik finds himself in a predicament that is familiar to us. He is trying to decide whether or not to quit a job, what type of education to pursue next, and, more generally, what life direction to take. Most of us have confronted questions like these. The familiarity of such questions sets the tone for this chapter, which will investigate questions of identity that are significant practical issues for real, presently living human beings. The question "What am I?" seldom arises, except among the very philosophical. The question "Who am I?" is more common. It *might* raise the issue of numerical identity but, if someone asks the latter question in earnest, she probably suffers from amnesia or another mental disturbance. The more ordinary sense of "Who am I?" inquires about one's identity in a familiar sense of the term that we may call *narrative identity*. Such related questions as "Who shall I become?" or "In what direction should I take my life?" ask about what we may call *self-creation.*

 The remainder of this chapter begins by picking up a question that was only partially addressed in Chapter 2: "What matters in survival?" A more complete answer highlights that we human persons care a great deal about our narrative identity – who we are, in the familiar sense of these words – and about our self-creation, or what we make of ourselves. Thus, the next section takes up narrative identity, formulating a cluster of questions connected with this theme, providing a conceptual framework for addressing such questions, and noting the close connection between narrative identity and several practical concerns discussed in Chapter 2. The section that follows explores self-creation, explaining what it is and addressing these questions: To what extent is self-creation possible? How does it relate to autonomy, self-narratives, identification, identity, and one's own values? (Because self-creation and autonomy prove to be closely related, the discussion includes a detailed exploration of the nature and possibility of autonomy.) Do demands of authenticity set moral limits on self-creation? The chapter concludes by tying numerical identity and narrative identity together into a unified conception of human persons.

WHAT ELSE MATTERS IN SURVIVAL?

Chapter 2 supported the thesis that numerical identity is necessary for what primarily matters, prudentially, in survival: In order to benefit in the future, one has to be around. A possible exception to this rule was noted in the case of cerebrum transplantation, but for all cases that we have encountered in real life thus far, identity was found to be necessary for what matters in survival.

But, while identity is necessary for what prudentially matters, for most of us it isn't sufficient. Some people, perhaps for religious reasons, value life itself – biological life – so much that they consider being alive not only necessary, but also sufficient, for what matters. "Life," they may say, "is a precious gift from God, and all I want for myself is to hold on to this gift as long as possible." But this attitude is very rare. Even among those who believe that the gift of life should not be taken away through human actions such as suicide or terminating life supports, few would agree that just continuing to be alive suffices for what *prudentially* (as opposed to morally) matters.

Nearly everyone wants more than maintaining numerical identity, or merely surviving. With few exceptions (which I will hereafter ignore), we human persons want, at a minimum, to retain the capacity for consciousness – to continue to be able to have experiences. But we also want to avoid a terribly low quality of life, or quality of experiences, so we would prefer death to survival with extremely poor experiential welfare and no prospects for improvement. (These points, remember, concern what matters prudentially; from an ethical standpoint, some will not prefer death, given this choice, because they consider unethical the only means to death.)

So far, our answer to the question of what else – besides survival itself – matters in survival has considered only *experience.* But to stress experience is to stress a relatively passive side of human persons: what we take in through the senses and process with our minds. Of course, we human persons are also *agents* – beings who act, sometimes spontaneously, sometimes after deliberation and planning. Agency seems no less central to what we are (at least during our existence as persons), and what we care about, than experience is.[1] So we want to retain not only the capacity for consciousness but also the capacity for action.

---

[1] Christine Korsgaard explores the importance of agency to personal identity, although I cannot tell whether she is addressing numerical identity, narrative identity, or some merging of the two ("Personal Identity and the Unity of Agency: A Kantian Response to Parfit," *Philosophy and Public Affairs* 18 [1989]: 103–31).

But that's not all we want. Suppose one is presented with a choice between dying soon or surviving with the capacity to experience and act but in a severely demented state. One is further told that this state is likely to be reasonably pleasant, but involves such severe memory loss that one would be unable to remember the previous day, and such severe destruction of executive functions that one could not plan more than a few minutes into the future. Some people would prefer to live in such a state rather than die, while others would prefer to die. But we would all prefer, over both options, a state of existence that permitted a reasonable degree of psychological continuity between different stages of our existence.[2] From the standpoint of the present, we would like to be able to *anticipate* having experiences and performing actions, not just accept the promise that we will, in fact, experience and act. Put another way, we want to be able to *identify* with the future subject-agent, regarding her subjectivity as a continuation of our own.[3] While some of us would assign *some* value to a pleasant life lacking such psychological continuity, we could agree that the latter is *a major part* of what matters in survival.

Another way to capture how we value psychological continuity is to think in terms of our self-narratives or inner stories. Each of us has a mental autobiography, an extremely detailed story of what we have experienced and done and a perhaps less detailed account of what we intend, or at least hope, to experience and do. This autobiography is not a mere listing of personal events and intentions. The story is richly colored by a sense of one's own beliefs, desires, values, and character – which affect which events are remembered and how they are remembered, make sense of and even help determine plans for the future, and shape the overall self-conception of an enduring protagonist. People differ greatly in how explicit their self-narratives are. Highly introspective people may frequently think through large segments of their inner story and even share those chapters with others. Many other people have a more implicit inner story – a set of memories, intentions, values, and other mental states that add up to a self-narrative that can be made explicit upon prompting

---

[2] Thus I suggest that the psychological view of Parfit and others is closer to the mark in addressing what matters in survival than in offering a theory of our (numerical) identity. But I also think some leading versions of this view underestimate the importance of identity as a *necessary* condition for what matters.

[3] Raymond Martin develops similar claims about anticipation and identification, but holds that numerical identity is not necessary to what matters in survival (*Self-Concern: An Experiential Approach to What Matters in Survival* [Cambridge: Cambridge University Press, 1998], ch. 3). We will examine identification further in Chapter 5.

from others.[4] Despite individual variations, every person has a mental autobiography.

Such self-stories are deeply connected with what we value in survival. Marya Schechtman argues, rather plausibly, that persons are precisely beings who create self-narratives.[5] Further, persons care deeply about continuing their self-narratives, which is to say continuing to exist as persons.[6] Her remarks suggest that we regard surviving as persons (as opposed to merely sentient beings, say) as intrinsically valuable. But surviving as persons may have instrumental value as well. As Jonathan Glover states, "[o]ur inner story lets us get our bearings as we act. Without it, all decisions would be like steering at sea without a map or compass."[7] If you lost your inner story, due to dementia or a brain injury, you might not know what to do since your sense of yourself and your values would be missing. This possibility suggests the instrumental value of having an inner story. Admittedly, another possibility is that after losing your inner story you would simply follow your immediate desires, displaying no paralysis or confusion about how to proceed. But, we who *as persons* contemplate such a state – following desires without any values or sense of oneself with which to evaluate and adjudicate among them – consider it far less preferable, other things equal, than existence as a person, as someone who receives guidance from an inner story. This judgment confirms the earlier claim that we regard having self-narratives as intrinsically valuable.

Our reflections suggest that what we value in survival isn't just survival per se. We value survival with the capacities for action and experience. But not the mere capacities; we want to continue to act and to experience. Further, we want our present self-narratives to continue to unfold and include the future actions and experiences, maintaining psychological continuity between ourselves now and ourselves later. But this isn't all.

---

[4] Although one's self-narrative might be significantly implicit, it could not be entirely hidden from one. Otherwise, it would not in any meaningful way be *that individual's* self-story. Cf. Marya Schechtman, *The Constitution of Selves* (Ithica, NY: Cornell University Press, 1996), pp. 114–19.

[5] Ibid., pp. 99–105. Somewhat similarly, Alasdair MacIntyre argues that we are essentially story-telling animals (*After Virtue*, 2nd ed. [Notre Dame, IN: University of Notre Dame Press, 1984], ch. 15). In saying that I find Shechtman's thesis plausible, I mean that I think it's *roughly* correct. As argued in Chapter 1, I don't think any *specific* analysis of personhood is authoritative.

[6] *The Constitution of Selves*, pp. 150–4

[7] *I: The Philosophy and Psychology of Personal Identity* (London: Penguin, 1988), p. 152. Cf. MacIntyre, *After Virtue*, p. 216.

For suppose one continued to live the life of a person but found, on one's deathbed, that the whole thing seemed a flop. A disappointing or unsatisfying life is not a flourishing life. Those of us whose basic needs are met, giving us the leisure to dream a little and a modicum of control over our futures, want *to make something* of our lives. For us, survival's value is partly instrumental: Continuing to exist as persons allows us not only to get our narrative-entrenched bearings when we act, as just explained, but also to pursue longer-term projects that we value and to become the sorts of people we want to be. Thus, for those who are fortunate enough to entertain such possibilities, much of what matters in survival is its making possible *projects of self-creation.*[8]

Since much of what we value in survival concerns our self-narratives and self-creation, these topics merit fuller exploration. It turns out that they have a great deal to do with identity. A self-narrative can answer the question "Who am I?" as this question is most commonly asked. The answer provides the person with her narrative identity. But who I am has a great deal to do with who I will become if I take an active role in shaping my future. Thus projects of self-creation flow from narrative identity and, as they do so, continue to write and often edit the narratives from which they flow.

NARRATIVE IDENTITY

Chapter 2 investigated our numerical identity, the question of what makes someone considered at a particular time one and the same individual as someone considered at a different time. Most people, in ordinary contexts, are not very interested in numerical identity. Certainly Erik the

---

[8] Glover develops this thesis in a discussion both forceful and pithy (*I*, p. 106). See also Richard Rorty, *Contingency, Irony, and Solidarity* (Cambridge: Cambridge University Press, 1989), ch. 5. Is the importance we assign to self-creation limited to the modern world or a certain stage of society? When, in earlier times, people's social roles were largely determined independently of their choices, would such agents have cared so much about self-creation? They may have cared less about it, in which case the importance given to self-creation may be a modern phenomenon. Do people in contemporary nonliberal societies, in which people's roles are often externally determined, care much about self-creation? Although I cannot defend my claim here, I think such people tend to care about self-creation – at least when their basic needs are met, giving them the opportunity to entertain such possibilities, *and they are aware of the possibility of self-creation.* That is, they are at least disposed to care about self-creation. Some evidence for my claim can be found in Martha Nussbaum, *Women and Human Development: The Capabilities Approach* (Cambridge: Cambridge University Press, 2000).

paralegal isn't. He is interested in his narrative identity. As Schechtman puts it, this sense of identity raises *the characterization question*: Which actions, experiences, values, and character traits can be ascribed to a particular person? Which of these characteristics make her the person she really is? This is the sense of identity at issue when someone has an identity crisis.[9]

Erik's predicament is precisely that of an identity crisis: "Who the hell am I, anyway?" We might imagine him following up with these questions: "What am I like – what sort of person am I? What are my central qualities? What's most important to me, giving me my sense of self? With whom or what do I identify?" Considering the impact of his parents on his sense of direction thus far, Erik might also ask, "How do other people shape my identity?"

A Framework for Understanding Narrative Identity

Let us consider a framework for addressing such questions. When someone like Erik raises the characterization question about himself – "Who am I?" – a helpful response will take this form: *"You are the individual who is realistically described in your self-narrative or inner story."* Several aspects of this response merit comment.

First, the response states "You are the *individual.* . . ." Although only a person will raise the characterization question with regard to herself, her inner story can include episodes that took place or will take place at times when the protagonist is not a person. Thus, one can meaningfully say, "I was born at such-and-such hospital," and "If I permanently lose the ability to remember my life history, don't keep me on life supports." It doesn't matter that one can't remember being born and might have trouble anticipating a state of severe dementia. One knows on the basis of others' testimony and everyday biological and medical knowledge that one was born and might someday become demented.[10] Thus the past event is appropriated into one's inner story, and the possible future state is appropriated as a possible continuation of the story.

---

9 *The Constitution of Selves*, p. 74. MacIntyre also stresses the role of narratives in exploring personal identity (*After Virtue*, ch. 15). See also Charles Taylor, *Sources of the Self* (Cambridge, MA: Harvard University Press, 1989), chs. 1 and 2.

10 At least we ordinarily take ourselves to know such things. If the biological view of our numerical identity is correct, as argued in Chapter 2, then considerations of numerical identity do not cast doubt on such claims.

Why does our general answer stated previously to the question "Who am I?" make one's self-narrative authoritative? Why not say that one is the person who has objectively had such-and-such a life and has such-and-such prospects for the future? Or that one is the person who is (realistically) described by other individuals? Why favor the first-person perspective, which can be distorted by self-deception, over objective or intersubjective third-person perspectives?

The reason is that only an answer that favors the first-person standpoint does justice to such a first-person question. If someone asked "Who am I?" with the reidentification question in mind – say, after a bicycle accident that resulted in amnesia – an objective, third-person answer would be germane: "You are Dan Scribner, born to Joseph and Ethel Scribner on June 20, 1939. You live at. . . ." But when one asks "Who am I?" in the more familiar sense of the question, one seeks a highly personal answer that, among other things, filters through objective facts about oneself, deeming only some of them salient. When Mr. Scribner's faculties function normally, that he is a husband and father is likely very salient to him, very much part of his (narrative) identity; that his fingerprints look such-and-such a way is probably not salient to him and not part of his identity, even if they identify him uniquely.

Nor is his identity comprised simply of those objective facts concerning him that he considers salient. *Interpretation* of the facts plays a major role in one's self-conception. Thus, being a jock, an opera buff, and a humane person may be part of his identity even if some people might reasonably disagree among themselves over whether he is all of those things. So long as a self-attribution is *within reason* factually, he can own the characteristic in question. Meanwhile, his neighbor might exercise and participate in sports no less than Scribner, yet being a jock is not part of *his* identity if he doesn't consider it important to who he is. Consider another example. Although Mr. Scribner once cheated on an exam in high school, he doesn't consider this act an important part of who he is; cheating was out of character for him, the deed was never repeated, and it had no significant effect on his personal development even if it occasioned a brief episode of guilt and repentance. Clearly, Mr. Scribner does not *identify* with his act of cheating. By contrast, another person might think his own single act of cheating is the tip of his evil dispositional iceberg, especially if he's often tempted to cheat. These examples suggest that a person's sense of herself is inextricably linked with who she is, making the first-person perspective indispensable to narrative identity. (In the next section, I will say more about why other individuals, who also subjectively sift through and interpret facts about a person in forming a sense of who she is, lack special authority in determining her identity.)

What if one's inner story is completely implausible? One might worry that we pay a high price for making the first-person perspective authoritative: allowing unrealistic or even delusional self-attributions to play a role in determining who someone is. I once met a woman who claimed that another woman had turned her into a snake. If we give a person's self-narrative free reign in determining who she is, then we seem obliged to conclude that this woman was once a snake! But the proper response to this sort of concern is not to abandon the first-person perspective as identity-constituting. The proper response is to insist that the self-narratives that qualify as identity-constituting are those that are *realistic* or *within reason*, given what we know about the person in question, about persons generally, and about the way the world works.[11] It is not realistic to suppose that a woman might have temporarily been a snake, so this part of the woman's inner story is not identity-constituting. "You are someone who was turned into a snake" is not part of a correct answer to her question "Who am I?"

What if someone's self-narrative is largely or (if this is conceivable) thoroughly unrealistic? Our framework states that one is the person who is *realistically* described in one's self-narrative. That means that one is the person described when we accept the bulk of one's inner story but not those parts that are clearly out of touch with reality. We can accept Mr. Scribner's self-attribution of being a jock, although some who know him well might not think of him this way. We could not accept his claim of being an opera buff if he had never listened to opera in his life and knew nothing about it. (Perhaps he is putting on airs in claiming to be an opera buff.) But unlike Mr. Scribner, Mr. Reilly is systematically deluded about himself due to extreme psychosis. If we reject those of his self-attributions that are way off base, we may have very little inner story remaining to serve as the basis for his narrative identity. Let's say he's right about who his family members are, where he attended school, and his age. But he also believes that his thoughts are controlled by the CIA through the Internet, that most of the people he sees on the street are spies, that animals are somehow secret agents, and that everything we see in the sky is an optical

---

[11] Schechtman offers essentially the same solution in an illuminating discussion that influenced this one (*The Constitution of Selves*, pp. 119–30).

illusion deliberately created by some international agency. In what does his narrative identity consist?

I suggest that, in the face of extreme tension between someone's first-person perspective and what we know about the world, we retain the first-person perspective but qualify its objects. After all, the narrative we seek in response to his asking the characterization question must, in some way, be *his* narrative; it has to be an *inner* story, not some external story primarily determined by others. So who is Mr. Reilly? He is someone who, for example, is *X* years old, has such-and-such family, and went to these schools. Continuing our answer, we employ qualifications: He is someone who *deeply believes* that.... This sort of reply to the characterization question in the case of a severely deluded person seems faithful to who that person really is without unpalatable metaphysical implications – such as that his thoughts really are controlled by the CIA through the Internet.[12]

The Role of Others in a Person's Narrative Identity

Our remarks about highly unrealistic self-portraits suggest one role other people play in determining someone's narrative identity. The knowledge possessed by nearly every person that human beings cannot become snakes prevents the woman's claim that she was turned into a snake from constituting part of her identity. Alternatively, if we allow the claim to play a role, then what counts as part of who she is merely the *belief* or *feeling* that she had been turned into a snake. So persons other than the narrator set limits on, or qualify, self-narratives that constitute identity. Note that it isn't only facts about the world – in this case, that human beings can't become snakes – that play a role here. Since self-narratives, like other sorts of narrative, are the sorts of thing that in principle can be shared with others, other people's *knowledge* of the relevant facts provides social reality checks that shape the story that will be accepted. (By contrast, as I will argue, other people's *distortion* of relevant facts carries no authority in shaping someone's identity if the protagonist dissents from the distorted claims.)

Should we further claim that others have as much authority in determining who someone is as the protagonist herself does? In an insightful

---

[12] Of course, we could also qualify the objects of the first-person perspective of someone like Mr. Scribner, saying that he is someone who *believes* that he is a jock, rather than simply accepting his claim that he is, as part of his identity. But, since his self-attribution, unlike some of Mr. Reilly's, is within reason, it seems unnecessary to make such qualifications in his case.

discussion, Hilde Lindemann Nelson suggests that narrative identity is a tissue connecting one's self-story with other people's stories about one.[13] But, as narrative identities involve *self-*conceptions, I find a privileging of the first-person standpoint the only reasonable option. In support of her view, Nelson states that "[m]y conception of myself as a skilled office manager who knows the firm from inside out goes nowhere if the new CEO thinks of me as the faithful old retainer who ought to be pensioned off."[14] True, in this case my self-conception will not lead to satisfying professional results. But that hardly vitiates the commonsense points that, regardless of what the CEO thinks, (1) I can *be* a skilled worker who knows the firm inside out and (2) if my self-esteem is healthy, I can *continue to regard myself as such*, making it part of my identity.[15]

Another way other people can affect someone's identity is by playing starring roles in her self-narrative. A large part of who you are is a function of your interpersonal relationships, some of which are central to your identity. For example, much of who you are might be described in terms of relationships with a life partner; your children, siblings, and parents; your closest friends; the neighborhood you grew up in; the schools you attended; your colleagues; familiar members of your religious community; and so on. Although these individuals, groups, and communities are not *literally* part of *you* – imagine what your weight would be if they were – they are certainly part of your *identity*. So, to some extent, their interests are your interests. That is why if my wife or daughter flourishes, I am ipso facto better off. It is not simply that their flourishing makes them better company, or easier to live with or care for. To the extent that they are part of my identity, our interests overlap and their well-being constitutes part of my well-being.

In addition to starring in one's inner story, and setting limits via reality checks on the story's content, other people can affect the tone and details through *mirroring*. Mirroring in this sense occurs when one person sees his own "reflection" in another person's apparent image, conception, or characterization of him. For example, if your friends and family frequently compliment you for (what they perceive to be) your integrity, you are much more likely to think of yourself as having integrity than if they didn't compliment you in this way. If nearly everyone you date comments

[13] *Damaged Identities, Narrative Repair* (Ithaca, NY: Cornell University Press, 2001), ch. 3
[14] Ibid., p. 81.
[15] Nelson eventually grants the first-person standpoint some priority over the third-person one in constituting one's identity (Ibid., pp. 104–5), but her supporting arguments are moral, whereas the issue of what constitutes narrative identity is primarily conceptual.

on how attractive (they think) your face is, this feedback is likely to affect your physical self-image. A great deal of mirroring, however, does not involve the ascription of a single trait, such as integrity or a handsome face, but rather the presentation of a highly complex, nuanced portrait – and the person mirrored may be quite unable to articulate in detail the reflections he saw or their effects on his self-image.

Suppose Beth, age forty, accepts an invitation to attend a party in her hometown, which she hasn't visited in many years. At the party Beth spends hours talking to old neighbors, friends, and fellow swim teamers, most of whom she has not seen since her teens. She finds the experience almost overwhelmingly stimulating and understands that much of her excitement involves seeing people from her past and, in them, a large part of her own past. She realizes that, besides being interesting and important to her in their own right, they also make her feel good about herself. As characters who played important parts in earlier chapters of her inner story, they are part of her identity. But at the party they also function as mirrors for her – though she can't describe with any specificity what they reflect back to her. Ruminating for several hours after the party, she feels somehow *consolidated*. Seeing all those old friends and acquaintances helps to shore up her sense of her life as a whole – as a narrative that hangs together and makes sense to its author – clarifying her past as part of who she is.

Of course, mirroring can have distorting effects, just as others can perceive one incorrectly. Suppose Julie has considerable mathematical ability, but her mother, believing women to be inherently inferior at quantitative thinking, has consistently conveyed to Julie that her interest in math is not worth pursuing. Julie grows up believing she is below average in math. When she gets high standardized test scores in quantitative reasoning, Julie considers them lucky. It is clearly false that she lacks mathematical ability, so being bad at math is not part of her identity. That she thinks of herself this way, however, is part of who she is (unfortunately), as is her intention not to pursue mathematical studies beyond high school. As we will see later, significant distortions in mirroring can undermine the autonomy of an individual's choices.

Narrative Identity's Close Fit with Certain Practical Concerns

Having identified a framework for understanding narrative identity, let us consider its fit with the practical concerns discussed at some length in Chapter 2. Although, in the world as we know it, *numerical* identity is necessary for what matters in survival and for the more specific practical concerns to get a foothold, it is typically not sufficient. On the other hand, if an individual persists over time – that is, maintains numerical identity – and a single self-narrative includes that individual at different points of time that we wish to consider, that is generally sufficient with respect to the practical concerns.

For example, although one can persist between two points of time without having self-knowledge over that time, if one also has a self-narrative that covers that span of time, one necessarily has *self-knowledge* spanning that time period. Similarly, to the extent that one includes oneself prospectively in one's self-narrative, one is capable of *planning and prudence* with respect to those future times. Moreover, having a narrative identity is nearly sufficient for moral agency and therefore *moral responsibility*; nearly everyone who has an ongoing self-narrative has sufficient decision-making capacity to qualify as a moral agent, as someone who can be morally responsible. Admittedly, there are exceptions. A normal two-year-old certainly has a (relatively simple) self-narrative, but it seems a stretch to attribute moral responsibility to her at this age. Indeed, developing a self-narrative may be a prerequisite to the self-control and grasp of reasons necessary for moral agency. Still, as these examples of the practical concerns suggest, narrative identity fits very well with these concerns. Perhaps it is only slightly inaccurate to generalize that numerical identity is necessary for these practical concerns to apply, while narrative identity (which presupposes numerical identity) is sufficient.

If we turn from the specific practical concerns to the broader question of what matters to us in survival, we find that narrative identity is, for many of us, clearly insufficient. We don't just want to *have* a self-narrative and know who we are. We want our inner story to go a certain way, and we want to *be* a certain way – to be a certain kind of person. These interests concern self-creation.

SELF-CREATION

Is It Possible?

As I use the term, *self-creation* refers to *the conscious, deliberate shaping of one's own personality, character, other significant traits (e.g., musical competence, athletic prowess), or life direction.*[16] When someone sincerely resolves to become more patient with other people, bolder in professional circles, or more accomplished at running, she intends to embark on a project of self-creation. When a person forms a conscious plan to become a successful stockbroker, and strives for years to meet this goal, she is engaged in self-creation; similarly with an individual who endeavors to become the sort of fully present parent that she didn't have while growing up. By contrast, if an individual just fills the roles set out for him, with no independent thinking about what to be and no conscious deliberation about how to reach his goals, he is not engaged in self-creation in the present sense. (Filling roles set out for one is compatible with self-creation, however, if the agent consciously and deliberately chooses that path with genuine appreciation of other possibilities.) As noted earlier, those whose basic needs are met and who have the opportunity to think about the possibilities of self-creation generally want to make something of themselves and their lives. Sadly, many people in the world, very possibly a majority, are too hungry, too economically deprived, or too socially oppressed to aspire to self-creation. For them, endeavors of self-creation either don't make it onto the mental radar screen or register as transient blips that disappear as quickly as most dreams do as we return to the business of waking life.

A Preliminary Challenge to the Possibility of Self-Creation and a Reply. But is self-creation possible even for those of us who are relatively advantaged? Some may doubt it, thinking that when we change, or pursue some life direction, the forces behind the "movement" are outside our agency. If you become a more cheerful person, such a skeptic might say that the change in you is due to a genetic predisposition, or the medication you took, or the social forces that pressured you to brighten your outlook or take that medication. If Erik decides to become a public defender, the skeptic might claim, his decision is determined by the combination of parental modeling in favor of law and the pain he experienced while working in a private firm, which caused a modest change of course. Self-creation and, indeed, free action generally are illusory.

---

[16] Cf. Glover, *I*, p. 131. The term *self-shaping* would have the advantage of not implying that your own efforts can literally bring you into existence, but *self-creation* is commonly used (and I like the way it sounds).

This skeptical view flies in the face of both phenomenology, which suggests to each of us (who is sufficiently fortunate) that she can change herself or her life direction to some extent, and everyday social observation, which suggests that other people sometimes manage such self-changes. From these commonsense perspectives, it seems that one may try, with some success, to become more disciplined; practicing disciplined acts tends to inculcate the virtue of discipline. One might even succeed in making oneself less disciplined if workaholism or perfectionism has come to feel unhealthy or too obliterating of life's joys. We may work at being more patient or more generous or more willing to stand up to authority, and sometimes we may change in the ways we want. We may aspire to orient ourselves more toward a relationship – or less, if we need to become more independent. When we succeed in making the changes we set out for, or in moving ourselves in the life direction we seek, it does not seem to us or other observers that what has happened is entirely due to factors outside our own agency. To be sure, our genetic makeup and our experiences, especially early experiences, have much to do with what is possible for us. But, within what is possible, our choices and efforts often play a significant role in determining what we do and become. Or at least that is the way it seems from an everyday commonsense standpoint (combining phenomenology and observations of others): Actively working on oneself and one's life can make a major difference to the results.

Before we consider an objection to the present appeal to phenomenology and social observation, it is worth emphasizing that neither suggests unlimited capacity for self-change and control over one's destiny. That we don't entirely control our destinies is almost too obvious to say, because the world outside our agency sets limits to what's possible. Al Gore, for example, could determine how hard to work in his campaign for president, but he could not single-handedly determine whether he would achieve his goal of becoming president; no one could *make* an electoral majority vote for him in a free election. Possibilities for self-change are also limited. People with addictions or obsessive-compulsive disorder know that their will isn't the only force driving their actions. It is widely appreciated that a character trait like laziness is not overcome in a single act of will; changes of disposition take time. We are frequently reminded that there are limits to what we can accomplish in changing our behaviors and characters, just as there are obvious limits to what our bodies can achieve in sports.

While active self-shaping is possible, it is only one crucial process that entails what we and our lives become. Possibilities for self-creation are limited by our enmeshment with other crucial factors and processes.[17] For example, we cannot escape the genetically encoded cycle of human life: the dependence of infancy and childhood, the turbulence of adolescence, the gradual loss of physical powers in advanced age, and so forth. Other critical factors concern the tools we are given to work with, especially our individual genetic endowment and the quality of our early environment. A final crucial factor involves the random, unexpected, yet momentous consequences of some of our choices. I once decided, somewhat reluctantly, to go to a Halloween party – where I happened to meet the woman who later became my wife and the mother of my child. According to the present perspective, while self-creation is possible, the range of possibilities available to a person is both opened up and limited by other major factors and processes that shape us and our lives.

The Deeper Worry: Hard Determinism. While acknowledging some limits to self-creation is helpful, a critic might reply, doesn't the present view beg the question of free will? If causal determinism is true, there is a strong case that, whatever phenomenology and social observation may *seem* to suggest, our actions are not really free – in which case self-creation is an illusion. This is a powerful challenge. While the relationship between determinism and freedom could easily occupy an entire chapter, or even a book, our discussion of this issue will be brief. Let us first sharpen the thesis of determinism and then consider its implications for freedom and self-creation.

According to a determinist, all events, including our intentional actions, are determined by causal laws. A superintelligent being who knew these causal laws and the state of the universe with perfect precision could, in principle, infallibly deduce what would happen in the future. In that case, any of your actions could have been predicted long before you were born. At first glance, this seems to rule out freedom of will (or action) and self-creation. If what you do is determined in advance, how can it really be up to you? If what you become is settled beforehand, how can the result be even partly of your own making?

A common response to this issue is that contemporary physics has undermined determinism. Quantum mechanics, in particular, suggests that events at the subatomic level do not display causal regularities. But the

---

[17] In developing the points that follow, I largely follow a discussion in Glover (*I*, p. 138).

prevailing theory of quantum mechanics does not so obviously threaten determinacy at more macro levels that include human action. Statistical regularity at the subatomic level may permit predictability at the level of brain events – neuronal firings – and this could allow us, in principle, to predict human actions.[18] Or if some set of factors other than brain events constitutes the subject matter of causal laws relevant to human behavior, they may occupy a level at which determinism holds. Let's assume, at least for purposes of argumentation, that determinism at the relevant level is true. What follows for freedom and self-creation?

One can imagine this response: "If determinism is true, then I'm not responsible for anything. In fact, nothing I do can affect anything, since all that happens is determined in advance. That, by the way, is why I didn't come to class this morning, Professor." However common this sort of response may be, it is sophomoric and confused. If determinism is true, then all that happens is determined by prior causes, but there is no reason to think that one's own intentions, choices, and efforts cannot be among the causes that determine one's actions. In ordinary circumstances, intending to come to class and trying to do so at the relevant time will get one to class. (Moreover, if our interlocutor were correct, I couldn't be responsible for blaming him.)

So, even if determinism is true, our agency plays a role in causal processes and has effects in the world. This is the beginning of a defense of the possibility of self-creation that falls under the heading of *soft determinism* – the view that although determinism is true, determinism and human freedom are compatible. (Hard determinism holds that they are not compatible, so the truth of determinism precludes freedom.) But, even if our choices make a difference in the world, a problem remains. To put it somewhat crudely, do we have any choice about our choices?

Suppose I leave my house this morning to go to work. I leave the house because I want to go to work, I believe that leaving the house is the only way I can do so, and nothing prevents me from leaving.[19] From an everyday, commonsense perspective, if, as in this example, I do what I want to do, then I act freely.

Now suppose, changing the example, that my department chair expected me to attend a meeting at work but I didn't show up. Did

---

[18] Ibid., p. 181.

[19] For an influential account of intentional action according to which the latter is the product of appropriately related beliefs and desires, see Donald Davidson, *Essays on Actions and Events* (Chicago: University of Chicago Press, 1980).

something prevent me? No. Did I mistakenly believe the meeting was to be held another day? No. If something prevented me, the chair would have excused me; if I had been confused about the date, he would have tempered his annoyance with me (perhaps holding me partly responsible for not trying harder to record the meeting date correctly). But, in fact, I didn't show because I didn't want to attend. The chair is furious with me, holding me fully responsible. But why is the desire component of my intentional behavior treated differently from the capability and belief components? Probably for two reasons. First, our desires reflect our more general motivations and attitudes, and, as Glover notes, our motivations and attitudes are central to our relationships with people in a way that abilities and relevant beliefs (e.g., "The meeting is today) generally are not.[20] Second, we generally assume that people can control their motivations sufficiently to do what we ordinarily expect people to do, such as show up for department meetings. Even if I didn't really feel like going to the meeting, I could, we generally assume, motivate myself sufficiently to get myself to the meeting. Thus, according to my chair, I am culpable for not attending because I could have made it. The bottom line in determining whether one is responsible for one's conduct, we tend to think, is whether one *could have done otherwise*. Since I could have come to the meeting had I chosen to, I am culpable.

But suppose I agree that I could have come *had I chosen to*, but since I didn't so choose, I couldn't possibly have come. (No one was going to carry me there kicking and screaming.) My chair, however, is unimpressed, insisting that I could have chosen to attend and, had I been properly motivated, I would have. But I might reply that I was not, in fact, properly motivated and in fact could not have been. Why not? Because, I might reply, ultimately all of my motivations and choices were determined in advance by causal laws. It is irrelevant what I might have done, in counterfactual situations in which I had different motives, made different choices, and had a different character. While we generally hold people responsible for their choices, motives, and character, it makes no sense to do so, according to the argument. Ultimately, in the actual world (as opposed to counterfactual situations), I could *not* have acted other than I did. Whether we do what others expect of us is, in the final deterministic analysis, a matter of luck, something beyond our control.

This imagined reply to the department chair expresses hard determinism. It challenges the ways we ordinarily think about freedom, arguing

[20] *I*, p. 185.

against its possibility. If the argument is correct, then self-creation would also appear to be impossible. Moreover, our everyday attitudes and practices of praise and blame, guilt and moral satisfaction would make no sense, resting on the illusion of genuine metaphysical freedom.

Soft Determinism and Frankfurt's Account. Confronted with this challenge to the possibility of freedom, the soft determinist must refine an account of freedom to argue persuasively for its compatibility with determinism. Perhaps the most influential of such accounts is that of Harry Frankfurt, who argues essentially as follows.[21]

Freedom of *action* (which, we may note, is often called *liberty*) is being able to do what you want. You want to do *X*, and nothing – such as external constraints on movement or others' coercion – prevents you from doing *X*, so you do *X*. But, in addition to wanting or desiring to act in certain ways, people often have attitudes or desires regarding such first-order wants or desires. He wants to go to the happy hour and get drunk, and does so, but wishes he weren't the sort of person who had such a desire; he wants not to have the desire to go to the happy hour. She wants revenge on her tormentor but wishes she could rid herself of this desire, which conflicts with her understanding of healthy psychology. Or she wants revenge and is glad she is sufficiently independent of liberal ideology that she can enjoy vengeful desires, which she wants to retain. Frankfurt calls the first-order desire that motivates the action we perform, the one that wins out if there are conflicting first-order desires, one's *will*. Thus, freedom of *will* is the ability to have the will one wants. The conflicted drunk is free to drink but lacks freedom of will, because his will, the first-order desire that prevails, is to drink, and that's not the will he wants. He *identifies* himself with his (first-order) desire to abstain from drinking through his second-order desire that this desire be his will. According to Frankfurt, someone who has both freedom of action, being able to do what she wants, and freedom of will has all the freedom that we could possibly want or imagine.

If only matters were so simple. Before examining challenges to Frankfurt's account, let us reconnect the discussion to our main themes. Our question is whether *autonomy* and self-creation are possible. What Frankfurt calls freedom of action and freedom of will are closely related to autonomy as philosophers normally understand this term. Freedom of action does not entail autonomy because a free action, like our alcoholic's

---

[21] "Freedom of the Will and the Concept of a Person," *Journal of Philosophy* 68 (1971): 5–20.

drinking (which he does because he wants to), can be performed without free will, stripping both the choice to drink and the drinking itself of autonomy. Both are driven by a force, his addiction, with which he does not identify. Let us say, as a first approximation, that autonomous action is action performed with free will, so that not only does one do what one wants to do; one wills what one wants to will. When we act autonomously, that is, we do what we want and want that first-order desire to prevail.

Naturally, the concept of autonomy applies not only to actions and choices, but also to communities, persons, and entire ways of living – the last connecting with the idea of self-creation. Our issue regarding self-creation is whether we can autonomously construct the lives we want to lead and autonomously become the persons we want to be. Since such constructing and becoming would consist in innumerable choices and actions, the possibility of self-creation depends on the possibility of autonomous choice and action. Since autonomous choice is implicit in autonomous action, I will focus on the latter. Is autonomous action possible?

This question is closely connected with that of whether Frankfurt's account, or a similar one, can provide a plausible conception of autonomy. That's because some leading doubts about whether his account adequately characterizes autonomous action (or freedom of will) prove to be doubts that our actions can really be autonomous in a deterministic universe. Let us therefore consider five leading challenges to Frankfurt's and similar accounts.[22]

Challenges to This and Similar Accounts. Challenge 1: This account wrongly implies that coerced action can be autonomous.[23] If you are held up at gunpoint and turn over your cash, you are unlikely to have a second-order desire that conflicts with your desire to fork over the money and save your skin. Even on subsequent reflection, you are likely to approve of your sensibleness (and the first-order desire). But since in this situation your second-order desire – that your first-order desire to obey the thug prevail – and your first-order desire are both satisfied, the present account implies, implausibly, that your action is autonomous.

---

[22] The following discussion of five criticisms and my reply to them draw somewhat from my "Autonomous Action and Autonomy-Subverting Psychiatric Conditions," *Journal of Medicine and Philosophy* 19 (1994), pp. 284–7.

[23] See, e.g., Irving Thalberg, "Hierarchical Analyses of Unfree Action," *Canadian Journal of Philosophy* 8 (1978), pp. 212–17.

Challenge 2: Since second- or even higher-order desires are not self-validating, they cannot be the key to autonomy. We act autonomously only when our actions are governed by our *values* – desires one has for things believed to be good – and not by mere appetites or conditioned desires. Higher-order desires are no less susceptible to conditioning than lower-order desires are (as examples will show in a moment).[24]

Challenge 3: The appeal to higher-order desires is insufficient for another reason.[25] Imagine that seventy-year-old June, a traditional housewife, cleans house, does the laundry, makes meals, and scrubs the dishes for her retired husband, who hires out his former areas of responsibility (e.g., yardwork) and spends his time watching television, surfing the Internet, and napping. Why doesn't June "retire" and hire out her traditionally assumed housework? She wants to do this work. Moreover, she identifies with her desire to do this work and, when feeling tired or lazy, she wants her desire to do housework to prevail over conflicting desires to slack off or visit friends. But suppose June's attitude about her domestic role is not the product of critical, independent reflection about various options open to her. Rather, she was virtually brainwashed from an early age to think of women as worthy only if they cheerfully performed such tasks. Her dominating husband reinforces this traditional image at every opportunity. In short, June is a subordinated, subservient housewife whose choices and actions hardly seem autonomous. Yet the present account implies that they are.

Challenge 4: The top-down structure of the model is arbitrary. According to the model, one who had acted nonautonomously in virtue of certain prevailing first-order desires (e.g., to get drunk) can achieve autonomy by revising them so that they accord with higher-order desires (e.g., to be free from alcoholic temptation); in such cases the higher-order desires motivate the change. But sometimes first-order desires (e.g., *not* to vacuum the house) initiate the revision of desires (e.g., of a higher-order desire to be a dutiful housewife) that leads to greater autonomy.[26]

Challenge 5: On the present account, one acts autonomously if one *identifies* with the desire that moves one to act as one does. But identification itself is a kind of action, raising the question of whether one's

---

[24] Cf. Gary Watson, "Free Agency," *Journal of Philosophy* 72 (1975): 205–20.
[25] See, e.g., Susan Wolf, "Sanity and the Metaphysics of Responsibility," in Ferdinand Shoeman (ed.), *Responsibility, Character, and the Emotions* (Cambridge: Cambridge University Press, 1988): 46–62.
[26] Marilyn Friedman, "Autonomy and the Split-Level Self," *Southern Journal of Philosophy* 24 (1986), pp. 30–2.

identification is autonomous. (Those who prefer to speak in terms of autonomous desires, as opposed to actions, can say that the issue is whether one's second-order desires are autonomous.) Suppose, for example, that Jaime, a scholar, has a very powerful desire to work. He works not just a typical workweek, but also on evenings and weekends. He works most holidays and takes vacations sparingly so that he can get more work done. He also experiences no conflict over his lifestyle. When he reflects on his behavior, he justifies his constantly working (and the desire to work) by reference to the valuable things that hard work makes possible – such as income, fame, and career opportunities – and the assumption that diligence and discipline are inherently admirable. So Jaime has a second-order desire to want to work, which harmonizes with his constant desire to work. While the present account suggests that Jaime works autonomously, this is not obvious.

Is his identification with the desire to work itself autonomous? There seem to be two possibilities, both of which pose problems for the present account.[27] One possibility is that his act of identification is autonomous. But, then, what makes it so? Perhaps he has carefully reflected on the value of the products of work, as well as on his assumptions about what's admirable, and has validated these judgments in light of this reflection. But the problem of autonomy now moves to this higher-order evaluation: Was it autonomously carried out? Because we may raise similar questions at any level, we seem to encounter an infinite regress.[28] Another possibility is that the acts of identification are *not* autonomous. Perhaps Jaime was heavily influenced by his parents, community members, and professional peers and never seriously questioned the assumptions behind the strong work ethic he now embraces. In that case, while there is no regress, it seems doubtful that one's actions can be rendered autonomous by a process – consisting of acts of reflection and identification – that is not autonomous. Note that the first possibility will likely fall into the second. An infinite regress of autonomous reflection and identification seems impossible for any finite being, so presumably such higher-order evaluation must stop with some values and desires that are simply "given" to one by factors outside one's agency – such as socialization,

---

[27] The two possibilities are nicely laid out in John Christman (although he doesn't ultimately endorse the critique), "Introduction," in Christman (ed.), *The Inner Citadel* (New York: Oxford University Press, 1989), pp. 10–11.

[28] See, e.g., Watson, "Free Agency," p. 218.

conditioning, and one's genetic makeup.[29] We have returned, in effect, to the thesis that determinism precludes autonomy and, with it, self-creation.

A Reply to These Challenges. Let us begin with Challenge 5's specter of infinite regress. Anticipating this concern, and thinking of someone like Jaime, Frankfurt responds as follows:

> There is no theoretical limit to the length of the series of desires of higher and higher orders; nothing except common sense and, perhaps, a saving fatigue prevents an individual from obsessively refusing to identify himself with any of his desires until he forms a desire of the next higher order.... When a person identifies himself *decisively* with one of his first-order desires, this commitment "resounds" throughout the potentially endless array of higher orders. Consider a person who, without reservation or conflict, wants to be motivated by the desire to concentrate on his work. He can properly insist that this question concerning a third-order desire does not arise.... The decisiveness of the commitment he has made means that he has decided that no further question about his second-order volition, at any higher order, remains to be asked.[30]

Let's say that Jaime decisively identifies with his first-order desire to work, thereby deciding, according to Frankfurt, that no higher-order issue about his second-order desire remains. Is that sufficient for autonomy?

Consider the specification of Jaime's case in which his work ethic – which establishes, or includes, the second-order desire to want to want to work – was largely handed down to him without his ever questioning it. I could imagine Frankfurt and critics reasonably debating whether Jaime works autonomously in this scenario. But suppose, changing the example, that Jaime used to embrace more of a leisure ethic. He now decisively identifies with his desire to work only because a hypnotist manipulated him, without prior consent, while he was hypnotized, causing him to make this decisive identification after returning to normal consciousness.[31] In this case, Frankfurt's view seems incorrect in implying that Jaime's acting in accordance with his work ethic is autonomous. After all, his identification with his desire to work is itself nonautonomous.

But maybe there is a way to avoid an infinite regress without implying that a relevant act of identification is nonautonomous. Perhaps the act of identification that is supposed to cut off further questions about

---

[29] Cf. Glover, *I*, pp. 186–7.
[30] "Freedom of the Will and the Concept of a Person," p. 16.
[31] Christman, too, cites a hypnotist in criticizing Frankfurt's theory ("Introduction," p. 10).

autonomy – thereby preventing an infinite regress – can be autonomous *in a different way from the way in which first-order desires and the actions that flow from them can be autonomous.*[32] We can allow, first, that since first-order desires may conflict and sometimes cause us to act nonautonomously, a first-order desire expressed in an action needs to be validated, through identification, for our action to be autonomous. But, second, our act of identification might be validated in a different way that neither suggests that the original problem of autonomy has moved to the higher level nor invites an infinite regress.

We need a condition for acts of identification to count as conferring autonomy on first-order desires, a condition that will clearly not have been satisfied in the hypnosis case. Gerald Dworkin suggests that one acts autonomously when one acts "for reasons [one] doesn't mind acting from."[33] But this is insufficient if one *would* mind acting for certain reasons if one gave them any thought. In a later writing, Dworkin mentions the need to distinguish ways of influencing one's reflective capacities that promote and enhance them from those that subvert such faculties. Where a person or action is autonomous, "identification is not itself influenced in ways which make the process of identification in some way alien to the individual."[34] If alienating influences are absent, then the condition of *procedural independence* is satisfied. (Dworkin analyzes autonomy as authenticity – higher-order identification with one's first-order desires – plus procedural independence.)

If procedural independence can make an act of identification autonomous, of what sorts of influence must identification be independent to qualify as autonomous? John Christman has a very promising suggestion: "[A]ny factor affecting some agent's acts of reflection and identification is 'illegitimate' if the agent would be moved to revise the desire so affected, were she aware of that factor's presence and influence."[35] This suggestion permits the agent, on learning of a factor's influence (e.g., that of socialization), to determine whether it is legitimate and consistent with autonomy or illegitimate. I believe this brings us close to an adequate account.

Suppose that, after much reflection and discussion with people he trusts, Jaime has come to understand how his work ethic – and therefore

---

[32] I learned of this ingenious move from Christman, "Introduction," p. 11.
[33] "Acting Freely," *Nous* 4 (1970), p. 381.
[34] "The Concept of Autonomy," in Christman, *The Inner Citadel*, p. 61.
[35] "Autonomy: A Defense of the Split-Level Self," *Southern Journal of Philosophy* 25 (1987), pp. 290–1.

his identification with the desire to work – was largely shaped by his family's influence. If this influence strikes him as alienating, so that he now feels he must reevaluate his work ethic, then the family influence counts as precluding, or at least diminishing, autonomy. If, on the other hand, he considers this influence fortunate, and the work ethic well worth having among the various possible ethics one might choose, then his family influence will count as consistent with autonomy.

But suppose Jaime, like many people, never achieves this degree of insight about the formation of his values and never seriously questions them. Would such an absence of critical evaluation strip his actions and way of life of autonomy? If so, then many and perhaps most people would rarely act autonomously (although they would often act freely, doing what they want to do). Even the most reflective people perform many humdrum, everyday actions – such as tying their shoes or talking to someone – without second-order reflection, yet these actions seem autonomous. But, in these situations, the agent usually *would* identify with the relevant first-order desires if the issue of their desirability were raised.[36] So a plausible approach is to permit *dispositional* identification and acceptance of influences to count as autonomy-conferring.

In that case, *would* Jaime continue to identify with his desire to work on recognizing the major influences that have shaped his attitudes and values? While in some cases there may be no uniquely correct answer to such a counterfactual question because the agent has no determinate disposition, often there will be an answer based on the agent's overall psychology. Thus, given one's psychology, one's identification with a first-order desire counts as autonomous in the absence of illegitimate influences, where the latter are those that one regards, or would regard, as significantly alienating; one's identification will not count as autonomous

---

[36] But appealing to dispositions, a critic might reply, reveals that the present account, like all multitier accounts of autonomy, is overly intellectualized; people's real deliberations usually don't go to second- and higher-order levels. One might consider an alternative analysis that avoids multiple tiers of preferences, such as this: A acts autonomously if and only if A acts (1) intentionally, (2) with understanding, and (3) without controlling influences that determine the action (Tom Beauchamp, "The Moral Standing of Animals in Medical Research," *Law, Medicine, and Health Care* 20 [1992], p. 12). But autonomous action involves a harmony between the motives that prevail in one's actions and one's more general attitudes and values – a feature that single-tier analyses don't satisfactorily capture. Note that, when a bird intentionally flies to her nest, she seems to satisfy the preceding three conditions no less than I do when I walk to my office. But birds lack the reflective capacities and self-control that are implicit in our concept of autonomy. Thus the alternative analysis apparently fails to provide sufficient conditions for autonomous actions. And so, I suggest, will any single-tier account.

if it was shaped by an illegitimate, alienating influence. This, then, is our criterion for autonomous identification.[37, 38]

Before returning to the five challenges, let us sharpen the analysis of autonomy: *A autonomously performs intentional action X if and only if (1) A does X because she prefers to do X, (2) A has this preference because she (at least dispositionally) identifies with and prefers to have it, and (3) this identification has not resulted primarily from influences that A would, on careful reflection, consider alienating.* Why substitute the language of *preference* for that of *desire?* While these two terms, as I understand them, refer to the same mental states, the connotations of *preference* are less potentially misleading in cases of

[37] Should our test for the legitimacy of an influencing factor ask about the agent's retrospective evaluation, as in the case or Jaime, or might retrospection come too late? Suppose Chiang has been thoroughly brainwashed by the party to accept its ideology. Consequently, he not only wants to burn books and torment intellectuals; he is even grateful for the brainwashing, which he calls "education." From his retrospective view, the party's influence is not alienating. Yet it is highly questionable whether his present party-line desires are autonomous, considering how they came about. Should we, as a test for an influence's legitimacy, ask *prospectively* whether it would be alienating? Perhaps prebrainwashing Chiang would consider such an influence illegitimate. If we hold that prospective hypothetical evaluation provides the proper test for an influence's legitimacy, that will plausibly imply that Chiang's book-burning desire is nonautonomous. But prospective judgment also has its limits. Suppose racist Jesse feels that liberal arts education is alienating, distorting people's minds so that they can no longer appreciate proper religious and parental authority. But his football scholarship, which he accepts in the hope of later playing professionally, leads him to receive such an education. He now wants racial equality and considers his old worldview narrow-minded and distorted. It's hard to believe here that prospective evaluation is the more reliable test of an influence's legitimacy. I suggest that we stick with retrospective evaluation as the appropriate test. If Chiang wants to burn books, identifies with his desire, understands the party's influence on him yet embraces this influence – not just publicly but in his heart – then Chiang has genuinely *changed* and his desire, I suggest, is autonomous, harmonizing as it does with his current worldview. It is his values, not ours, that (partly) constitute his autonomy. Another possible approach, however, would be to give prospective and retrospective evaluation equal weight in determining whether an influence is legitimate, bearing in mind that sometimes there will be no determinate answer.

[38] Is this criterion uncomfortably relativistic? Suppose Fritz wants to get seriously involved with Ursula and identifies with this desire. Suppose also that this identification is significantly influenced by an unconscious desire to be with his deceased mother. If Fritz is an uptight, nineteenth-century Viennese, he might be repelled by this influence on learning of it and reject it, so that he does not act autonomously in seeking serious involvement with Ursula. But if Fritz is a contemporary New York intellectual, he might laugh at this influence on discovering it – after all, Freud's ideas are hardly shocking anymore – and embrace it without embarrassment. In this case, his seeking a serious involvement with Ursula counts as autonomous. I wholeheartedly accept the implication of the present account that the same action with the same influence could be autonomous or not, depending on one's culture, socialization, and so on. After all, autonomy is partly a function of one's values, which are shaped by social factors.

conflicting desires where one doesn't "really want (desire) to" do what one does; one does, all things considered, prefer the act one intentionally performs.

Now for the challenges. Challenge 1 concerns coercion. If you hand over money to the thug, you do so because, in the circumstances, you prefer this action to any other possible action, such as fighting, fleeing, or cursing the thug. But, on our analysis, the action won't count as autonomous, on either of two ways to interpret the case. One possible interpretation is that you do not have this preference to hand over the money because you identified with and preferred this preference; you have the preference because someone threatens you. In this case, condition (2) is not satisfied. Another interpretation is that, on reflection, you would, considering the drastic circumstances, identify with and prefer the prudent preference to surrender the money – and this disposition is the reason you preferred to act this way. Then condition (2) is satisfied. But condition (3) is not, because you would clearly regard the influence that determined your second-order identification – namely, the threat of death – as alienating. So the revised account apparently provides the right answer in coercion cases.

According to Challenge 2, our actions are autonomous only if they flow from our values. Second-order desires, or preferences, may result from conditioning that throws their autonomy into doubt. But, assuming (consistent with this challenge) there is such a thing as autonomous action, surely not all forms of conditioning preclude it. Our account plausibly distinguishes cases where conditioning and other influences prevent autonomous second-order identification – and therefore autonomous action – and cases where they don't. Moreover, the requirements of reflective identification and acceptance of the influences that bear on the latter ensure that actions deemed autonomous will flow from one's value system. For one's values will provide the grounds for identification or rejection of one's motives for actions, and for acceptance or rejection of the relevant influences.

According to Challenge 3, the multitier approach to autonomy implies that a thoroughly subordinated housewife would act autonomously in carrying out her role. This is unlikely with our revised account. June may identify with her desire to perform her traditional duties. But, given the unfairness of her current situation and the way in which the men in her life have denied her any choice about her role, it is very unlikely that condition (3) would be satisfied. Most likely, she would consider alienating the influences that shaped the identification in question if they were brought to light along with other possible roles for women.[39]

Challenge 4 charged the top-down structure of the model with arbitrariness, noting that sometimes first-order desires (e.g., not to vacuum) catalyze changes of higher-order desires (e.g., to be a dutiful housewife) on the way to greater autonomy. But, in cases like these, the autonomy-seeking rebellion at the lower level suggests that some condition of autonomy had previously not been met. One possibility is that the agent had the lower-order desire (e.g., to vacuum dutifully) not because she preferred to have it, but because she was subject to alienating influences such as subordination. Alternatively, if she did prefer to have the desire and genuinely identified with it, perhaps *that* was due to alienating influences such as subordination or excessive socialization. Thus, while revisions of preferences may occur at any level in leading to autonomy, this doesn't cast doubt on the requirement of higher-order identification. What the model really requires for autonomy is a kind of reflective harmony in one's system of preferences and values, a harmony that can be achieved in different ways.

Challenge 5 posed a dilemma regarding the second-order acts of identification on which autonomous action depend: Either such identification must be autonomous, a requirement that seems to lead to an infinite regress of higher-order autonomy-conferring mental acts, or such identification is not autonomous, in which case it seems incapable of supporting autonomous action. Our discussion of Jaime, in light of a criterion for autonomous identification that does not seem to invite an infinite regress, has already addressed this challenge.

But a skeptic might reply as follows: "Your criterion for autonomous identification permits factors external to one's agency to affect, in crucial

---

[39] Accepting these influences, on careful reflection, seems tantamount to accepting the idea that men should treat women inequitably, that women are morally subordinate to men. Admittedly, a woman might retain this belief, even after coming to understand the effects of socialization and reflecting on alternative, more respectful ways of viewing women. Such a woman, however, strikes me as either morally obtuse or psychologically unhealthy. Since our account does not absolutely preclude that such a person may choose a subordinate lifestyle autonomously, we face a theoretical choice: Either accept this implication or add a condition that requires reasonable moral perceptiveness, healthy self-respect, or the like. Susan Wolf has suggested that some such condition is necessary ("Sanity and the Metaphysics of Responsibility"). But, as John Christman argues, such a requirement – he has in mind requirements of "external rationality" more generally – might strip autonomy of the very self-governance that lies at the center of the concept ("Introduction"). I leave this subtle issue open.

ways, one's disposition to identify with this or that desire. A person's genetic makeup and all the social forces that bear on him (regardless of whether he would consider them alienating) will crucially affect what Jaime or anyone else will be disposed to identify with. But these factors are outside one's control, making it incomprehensible why one's identifications, preferences, and the actions they motivate should ever count as autonomous. Even if Jaime's feverishly working meets the three proposed conditions for autonomous action, the dependence of his values and choices on factors outside his control, and the fact that (assuming determinism is true) he could *not* have acted otherwise, undermine all claims to autonomy. And as autonomy goes out the window, so does the possibility of self-creation. Both are impossible."

The first thing to notice about this argument is that it embraces skepticism about autonomy itself and not simply our analysis of autonomy. So it doesn't challenge the claim that, assuming there is such a thing as autonomous action, our analysis captures it fairly well. That leaves us with the issue with which we started: Assuming determinism is true, are autonomous action and self-creation possible? In a way, we have reached an impasse. I cannot convince everyone that autonomy and self-creation are possible when some people assume that these phenomena require complete independence from external influences. Instead, I simply advance two claims: (1) If human beings are capable of autonomy and self-creation, then this chapter's characterization of these phenomena is satisfactory and (2) there are good reasons to think that the skeptic's stringent requirement – that autonomous action and self-creation require complete independence from factors beyond an agent's control – is unreasonable. I have already defended assertion (1) at length. A few further remarks will have to suffice in defense of claim (2).

If an agent does *X* because he prefers *X* to alternative possible actions, he has this preference because he identifies with and prefers to have it (at least dispositionally), and if the external factors that influence his identification are not such that he considers them or would consider them alienating, then it is very clear that the agent is actively involved in his choice and action. Even if, at a deep metaphysical level, he could not have acted otherwise, his agency – his values, choices, and efforts – are a crucial part of the overall causal processes that yield the action. We might even say that the agent is causally determined to act autonomously![40] But

---

[40] Frankfurt makes essentially the same point ("Freedom of the Will and the Concept of a Person," p. 20). Cf. Glover, *I*, p. 130.

this, of course, assumes the soft determinist thesis that determinism and autonomy, or free will, are compatible. Given the impasse noted earlier, it is hard to avoid begging some questions.

But I can avoid question-begging in pointing out that the ideas of autonomous action and self-creation are not going away soon. In ordinary life, we assume that what we do can be more or less autonomous, and that the shape of our lives can be more or less affected by our own agency. Moreover, we assign great moral and prudential importance to the differences. To disrespect a competent adult's autonomy is morally problematic. To engage in self-creation is a large part of what we hope for in life if we are lucky enough to entertain such possibilities. Those who wish to participate (intelligibly) in the immense variety of practices that implicate the concepts of autonomy and/or self-creation must assume that the latter are possible.[41] The remainder of this book makes this assumption.

How Self-Creation Relates to Narrative Identity and Other Key Concepts

Having explored the meaning and possibility of self-creation – and its conceptual cousin, autonomy – let us reconnect self-creation with narrative identity and other concepts that have figured prominently in our discussion.

Narrative identity is one's sense of oneself as the protagonist in one's own life story or self-narrative. Your sense of yourself helps you decide what is worth doing and which of your characteristics are *really* yours. Self-creation, meanwhile, is the deliberate and conscious shaping of one's own characteristics and life direction. As noted early in this chapter, self-creation projects flow from narrative identity and, as they do so, continue to write and often edit the narratives from which they flow.

Should we say, then, that self-creation is the inevitable process of continuing to write one's own self-narrative? No, because self-creation, as we saw earlier, is not inevitable and is very difficult or impossible for some people due to their circumstances. Further, there may be people who are in a position to engage in self-creation, their circumstances being

---

[41] See P. F. Strawson, "Freedom and Resentment" (*Proceedings of the British Academy* 48 [1962]: 1–25), for a discussion of our common attitudes and practices that implicate the assumption of free will, or autonomy, and the question of whether such attitudes and practices could be foregone.

sufficiently favorable, but who just never pick up the reins and take control of the bucking bronco that is their life. Self-creation occurs when one doesn't just watch one's life story being written or look elsewhere as the chapters fly by; it occurs when an individual takes an active role in authoring the biography, making it a *lived* autobiography. Rather than not wondering about how later chapters will turn out, or wondering how they'll turn out but with no sense of controlling their direction, the self-creator endeavors to write those later chapters – and perhaps, in light of the evolving story, edit earlier chapters as different themes emerge as critical and self-discoveries put old details in a different light. So, where self-creation occurs, it flows from one's narrative identity or sense of self, and where the latter doesn't enable one to take an active role in self- and life-shaping (whether or not circumstances make these possible), self-creation will not occur.

Although we have already discussed the relationship of self-creation to autonomy in arguing for the possibility of both, a few further remarks may be illuminating. We have seen that autonomy can characterize not just ordinary actions and acts of identification, on which we focused, but also desires, choices, and individuals. Importantly, human beings can live their *lives* more or less autonomously. It makes sense, then, to think of self-creation very roughly as *autonomous writing of self-narratives.* Jasmine pursues a career in academia, say, because she wants to and because, in view of the options open to her, she reflectively identifies with this ambition. In this self-creation project, she takes an active role in guiding her life direction in accordance with her values.

Autonomy and self-creation connect in another interesting way. When we act autonomously, so that we identify with the desires that prevail in our actions, those prevailing desires tend to reflect our values. Not necessarily all our values, some of which may concern ethical issues that are remote from our everyday choices. The desires with which we identify in autonomous action reflect our values concerning what sort of life would be worthwhile *for us* and what sort of people we want to be.[42]

From another angle, we may say that someone engaged in self-creation *identifies* with the person she is becoming, or trying to become, and with her apparent life direction, or the one for which she is aiming. The relevant sense of *identification* here is evaluative; it is not that associated with numerical identity, since it is trivially true that one is numerically

---

[42] Cf. Glover, *I*, p. 130.

identical to the individual one will become. (Chapter 5 will examine several senses of identification.)

Do the Demands of Authenticity Set Moral Limits on Self-Creation?

We turn now to an issue about a possible limit to legitimate self-creation, an issue both interesting in its own right and important to our discussion of so-called enhancement technologies in Chapter 6. One sometimes hears the assertion that certain self-creation projects are morally suspect, or even unethical, because they are inauthentic.[43] Is this correct? As a preliminary, let's clarify the concept of authenticity.

Authenticity, in this context, may be understood as *being true to oneself and presenting oneself to others as one truly is.*[44] Presenting a false self – pretending to be someone one is not – is inauthentic. Such inauthenticity is evident in a preppy East Coast schoolboy who suddenly dresses down and adopts a southern accent in order to impress a girl from Louisiana. Another example is a middle-aged woman who talks endlessly about the 1960s as if she had been deeply involved in the era's social movements, although she was abroad at the time and didn't participate at all. Consider also a nouveau riche couple who, in order to fit in with and impress new associates, pretend to be old-money aristocracy – when in truth they retain many of the values and tastes associated with their humble origins. Sometimes inauthenticity involves communicating falsely with oneself, as in the case of a man who, ignoring relevant evidence, tells himself that he is a scientific genius (an example developed later). Authentic people, by contrast, express who they are through their choices and actions, without pretense or artifice. Because there is little tension between who they really are and the personas they present to themselves and to others, they often strike us as particularly natural and comfortable with themselves.[45]

---

[43] I have heard this charge mostly in conversation. Carl Elliott has examined at length the tension in American culture between the values of self-fulfillment (or self-creation) and authenticity. See his "The Tyranny of Happiness: Ethics and Cosmetic *Psychopharmacology*," in Erik Parens (ed.), *Enhancing Human Traits: Ethical and Social Implications* (Washington, DC: Georgetown University Press, 1998): 177–88 and *Better Than Well: American Medicine Meets the American Dream* (New York: Norton, 2003).

[44] This is my understanding of our shared concept of authenticity. For an influential discussion of authenticity as a moral ideal in the modern West, see Charles Taylor, *The Ethics of Authenticity* (Cambridge, MA: Harvard University Press, 1992).

[45] Arguably, authenticity is compatible with *some* forms of self-deception. One might be decidedly optimistic in viewing one's life and the challenges it presents in order to facilitate success, since optimism tends to be energizing. But perhaps (full) authenticity

Might some self-creation projects run afoul of authenticity in morally problematic ways? Let's assume that the nouveau riche couple, for example, *deliberately and self-consciously* set out to put on airs and play the false role of old-money aristocracy, so that their social transformation counts as a self-creation project. Surely, *to the extent that they are being dishonest to others*, their self-presentation is morally problematic. In general, intentionally presenting oneself falsely to others subverts our expectations for honesty and sincerity – and may involve actively lying, adding unambiguously wrong action to bad character. But these ethical concerns are explicable by appeal to widely embraced norms for virtue and conduct, leaving unclear whether the charge that a self-creation project is *inauthentic* adds any distinctive moral content.

Suppose, then, that someone's inauthenticity consists in being untrue to himself, without any dishonesty toward others. Imagine a college student who is socially without pretense, yet lies to himself in one important respect. He keeps telling himself he is a scientific genius – because he wants to believe this – despite *overwhelming* evidence to the contrary. He presents a false self to himself. And he begins to organize his life around this self-deception, researching top graduate programs in theoretical physics, taking as many science courses as he possibly can (getting C's and B's), purchasing basic laboratory equipment, and so on.[46] He even gets accepted by a graduate program – one desperate for tuition-paying students. As the years continue, he keeps organizing his inner story around the theme that he is a misunderstood genius whose weak public performances, such as mediocre grades, are misleading for one reason or another. So this case, too, is an instance of inauthentic self-creation.

Is it morally problematic? Well, is it unethical to deceive oneself? Presumably not, if the relevant mental acts are beyond our control – which is quite possible, assuming the self-manipulation occurs unconsciously. I am not sure even intentional, conscious self-deception is unethical. But,

would require the agent to recognize, at least some of the time, that this bias toward the bright side is a self-management strategy rather than a reliable indicator of the agent's warranted beliefs, as Patricia Greenspan suggested to me.

[46] Although self-deception is paradoxical – to lie, one must know the truth, but to be fooled, one must not know the truth – we all know that it occurs frequently. Apparently, self-deception is possible because we are capable of more or less compartmentalizing different parts of our cognitive world, preventing the idealized consistency and rationality of which an information-processing system is, in principle, capable. See, e.g., David Pears, *Motivated Irrationality* (Oxford: Clarendon, 1984), ch. 3.

for the sake of argument, let's assume that all self-deception is unethical or at least morally problematic.

But, then, the fact that this case involves *deception* – toward oneself – may fully explain why, as we are assuming, it is morally problematic. Once again, there is no reason to suppose the charge of inauthenticity per se adds any morally significant content. On the other hand, one might deny that what's morally troubling about inauthentic self-creation needs some basis other than the familiar one that deception is wrong and contrary to virtue. For one might claim that inauthenticity is *inherently* deceptive and for that reason morally problematic. Perhaps that's right. So let's grant that assumption, too, at least for the sake of argument. (I grant all these assumptions for the sake of argument, because doing so won't prevent me from undermining the conclusion toward which the present reasoning is heading.)

So do the demands of authenticity set moral limits on self-creation? Yes, at least insofar as self-creation should not be founded on deception. Now, the cases of deception we have considered are cases in which self-creators refuse to acknowledge – to others or themselves – who they are, what sorts of persons. But consider a subtly different class of cases that some people find morally troubling: those in which people want to *change themselves* in ways that might be considered untrue to themselves.

Consider three examples. (*A*) An unconfident, physically unimposing teenager wants to become more assertive, confident, and physically fit. He pushes himself to act differently in social settings, desensitizing himself to social fears; seeks pointers from "cool" neighbors who are a bit older; and begins regular exercise, including running, weight lifting, and basketball. Within a few years, he has shed the nerd image and feels in command of his social life. (*B*) Unlike the self-deceiver described earlier, this physics student understands that she is no genius. But she has moderate ability and wants to succeed in her major; she finds it enjoyable and values science as a constructive enterprise. Through consistent hard work and the gradual accumulation of genuine expertise, she exceeds expectations, eventually establishing herself as a research scientist who makes modest contributions to her field. (*C*) Tired of being told that she is a wonderful friend and conversationalist – remarks she hears when being rejected by men she wants to date – a plump, flat-chested young woman decides to become more physically attractive. She eats more healthfully, joins a gym and uses it regularly, and undergoes cosmetic surgery – reshaping her nose and enlarging her breasts. Soon she has a more active love life and finds that in professional circles her physical appearance gives her confidence and even a measure of power that she lacked before.

These three individuals have developed self-creation projects that involve deliberately changing themselves – more specifically, their bodies, personalities, and/or levels of professional competence.[47] Are these individuals, in changing themselves, being untrue to themselves, inviting a charge of inauthenticity? "You're not being honest about who you are," our imagined critic might say to any one of them. "Your self-creation project is false."

That this concern is at least partly misplaced is suggested by the likelihood that we will find at least one, very possibly two, and perhaps all three of the self-creation projects admirable. The easiest case is that of the physics student. Exceeding expectations, when one is honest with oneself and others, seems admirable, not morally suspect. In doing extremely well with her less than awesome natural endowment, the student does justice to the full range of her ability; she doesn't pretend to be something she's not.

The case of the teenage nerd who becomes more fit, confident, and assertive might give one pause if one doesn't share his enthusiasm for social grace and control. Perhaps one finds these values superficial and those who seek them overly competitive. But, assuming the teenager doesn't become obnoxiously aggressive, or reject others in becoming socially graceful, this sort of criticism seems excessive. And, even if he did become obnoxious and rejecting, the only clear moral fault would lie with his new attitudes and behavior. If one charged him with inauthenticity, stating, "He's not really like that," it would be appropriate to respond, "Look for yourself!"

I suggest that the only genuinely troubling case, morally, is that of the young woman who transforms herself into a sexier form. Those uncomfortable with this case are likely to feel especially skeptical about her decision to undergo cosmetic surgeries. These surgeries entail certain nontrivial physical risks. Even if we set aside these concerns by assuming, for the sake of argument, that the procedures are physically safe by anyone's standards, some will remain morally troubled. One might be disturbed

---

[47] Admittedly, these individuals may be somewhat unusual in the degree to which they actively take control of their lives. But some people do self-create in such ways. On a personal note, four times – at ages eight, nine, seventeen, and twenty-nine – I have made conscious decisions to change myself in important ways and restructured my behaviors accordingly. In the latter two cases, the changes were comparable to cases *A*, *B*, and *C* in degree of impact.

that this woman – whose self-creation project requires a great deal of effort, time, and expense (in sum, considerable cost to herself) – may be reflecting socialization into sexist norms for what constitutes a beautiful woman: slim, large-breasted, straight-nosed, and so on. Moreover, while vigorous exercise is generally admirable, one may be concerned about even this part of her self-creation project if the impetus to exercise is the internalization of sexist norms of beauty.

But, if the woman strongly desires the changes she seeks, is there any basis for saying that her goals are inauthentic? "Yes," one might say, "because her desire to look a certain way doesn't really come from *her*, being born of socialization in a sexist culture." Notice that this concern could also be expressed by one who questioned the *autonomy* of her desire to become slimmer, large-breasted, and so on.[48] And this hints at a promising way to address such cases, for in view of our earlier discussion of autonomy, the concern about autonomy is clearly legitimate.

I suggest that any self-creation project that is autonomous and honest is ipso facto authentic. Honesty is necessary because if, for example, one autonomously pursues a life course that involves systematic deception to others about who one really is, such a self-creation project would clearly be inauthentic. In the case under discussion, legitimate concerns about authenticity seem to be concerns about autonomy: Is her self-creation project really hers, or is she capitulating to social forces that largely determine her choice? The answer depends on the details. Suppose she is aware of sexist socialization and its impact on her desires, seriously considers options other than those of conforming to the sexist norms, but decides that, on balance, she really prefers to strive for the socially sanctioned type of beauty. In this scenario, she acts and chooses autonomously – and therefore, I am suggesting, authentically. Suppose, alternatively, that her desire to change herself in the ways described is largely a product of socialization, which she would consider alienating if only she understood its impact, and she would choose differently were she more perceptive about her psychological situation. In this scenario, her choices are neither autonomous nor authentic. She is not really in the driver's seat of her beauty-seeking behavior and the self-image she pursues is, in an important sense, not her own.

---

[48] Recall that Gerald Dworkin defines autonomy partly in terms of authenticity – which he understands as higher-order identification with the first-order desires that prevail in action – providing some confirmation for the intuitive sense that authenticity and autonomy are importantly linked.

But one might claim that details pertaining to honesty and autonomy in a given case are beside the point, because drastic self-transformation is *inherently* inauthentic: "Just in seeking to change oneself so drastically, one is being inauthentic, abandoning one's true self – either pretending it doesn't exist or denying its importance." While sensing that many people resonate with this sort of reasoning, I find it unpersuasive. It suggests, misleadingly, that the self is given as a largely fixed, unalterable entity. This static model of the self fails to recognize, or at least respect, the sort of self-creation in which we deliberately change *ourselves* – or, more specifically, our personality, character, or abilities – as opposed to merely our life direction. I see no good reason to object to self-change just because it involves major change. While some people are romantically attached to the way things are, including the way *they* are, many people are not. And it would disrespect the latter people's autonomy to suggest that their interest in improving themselves according to their own lights is inherently problematic. To say this interest is inauthentic begs the question of who they are – and *they*, not we, are the authors of their self-narratives.

On the present view, one can authentically change oneself, even radically, if one does so autonomously and honestly. If this is correct, then authenticity sets no moral demands on self-creation beyond those connected with honesty and autonomy. In Chapter 6, however, we will revisit the charge of inauthenticity and examine in depth a related charge: that some radical self-transformations violate morally inviolable core characteristics. The findings of this section are therefore tentative.

CONCLUSION: BRINGING THE TWO SENSES OF IDENTITY TOGETHER

In this chapter, we have explored the important human phenomena of narrative identity, self-creation, and several related themes such as autonomy and authenticity. We have found that when someone is engaged in self-creation – conscious, deliberate self-shaping – the latter flows from the individual's narrative identity and, as it does so, continues the life work of writing the inner story from which it flows. Narrative identity, in turn, is the sense of human identity that most concerns people in everyday life, the sense at issue when someone wonders what sort of person she is, what's most important to her, and with what or whom she identifies in securing her sense of self.

The theoretical work completed in this and the previous chapter can be succinctly summarized in the following way: Human persons are

(1) essentially human animals and (2) characteristically self-narrators and (where circumstances permit) self-creators who care about continuing as such. Thus, we – who are now human persons – are human animals, but not necessarily persons, throughout our existence. But we human persons all have, and will continue to have so long as we are persons, inner stories whose overall character and direction matter to us. More fundamentally, much of what matters to us is our continued existence *as persons* – as beings whose complex forms of consciousness make self-narration and self-creation possible. Yet we cannot continue to exist as persons unless we continue to exist. That is why narrative identity, on the present view, presupposes numerical identity.

4

Identity, What We Are, and the Definition
of Death

How should we understand human death? What role should personal identity theory play in determining an answer? After providing a historical, conceptual, and scientific introduction to the debate over the definition of death, this chapter addresses these and related questions.

INTRODUCTION TO THE DEBATE

Before the middle of the twentieth century, there was little controversy over the criteria for human death. Occasional cases of premature burial in the nineteenth century generated alarm, but concerns about the accuracy of diagnosing death largely abated by the turn of the century.[1] In the 1960s widespread dissemination of such new technologies as respirators, defibrillators, drugs that stimulate various bodily functions, and intensive care units created the possibility of radically separating cardiopulmonary failure and cessation of neurological function. Thus, respiration and circulation could be artificially maintained for months or even years after consciousness was irretrievably lost. These developments motivated the question of what constituted human death.

Various practical factors provided additional incentive to address this question. Resource constraints in hospitals raised concerns about expensive yet futile treatment. When was it legitimate to turn off life supports? Certainly, they could be discontinued for patients who were already dead.

[1] Ronald Cranford, "Criteria for Death," in Warren Reich (ed.), *Encyclopedia of Bioethics*, 2nd ed. (New York: Macmillan, 1995), p. 529.

Meanwhile, interest in transplantable organs created the incentive not to delay unnecessarily in determining that an individual had died. Organs, transplant surgeons realized, were more viable the fresher they were, but could not be removed from living patients; otherwise, the surgery itself would kill the donors, violating the law against homicide. In addition, growing mistrust of the medical profession – part of the 1960s' growing mistrust of authority figures and institutions generally – led many physicians providing care near the end of life to seek protection from unwarranted charges of homicide or other impropriety. Such protection would be possible only if professional bodies or the law provided clear guidance for the determination of death.[2]

Traditionally, death was understood as *the permanent cessation of cardiopulmonary function*. Thus, in determining whether someone had died, one would check for a pulse, moisture on a mirror held in front of the mouth, or other signs that the heart and lungs were working. Before the development of respirators and other modern life supports, cardiopulmonary function entailed that the brainstem continued to function. But respirators permitted continued respiration and circulation even when an individual's entire brain, including the brainstem, was nonfunctional. Was such an individual alive?

In 1968 the Ad Hoc Committee of the Harvard Medical School, led by physician Henry Beecher, issued a report that answered this question negatively.[3] Pragmatic, clinical, and devoid of philosophical argumentation, this report offered criteria by which physicians could determine that a patient had permanently lost all brain functions, from primitive brainstem reflexes to consciousness, and recommended that physicians use these clinical criteria in diagnosing death. Hence *the whole-brain standard*: human death as *the irreversible cessation of all functions of the entire brain, including the brainstem*.[4] Individuals who had met this standard were called *brain dead*. Interestingly, the committee did not recommend legal adoption

---

[2] For an illuminating discussion of the early years of the definition-of-death debate, see Martin Pernick, "Brain Death in a Cultural Context: The Reconstruction of Death, 1967–1981," in Stuart Youngner, Robert Arnold, and Renie Schapiro (eds.), *The Definition of Death: Contemporary Controversies* (Baltimore: Johns Hopkins University Press, 1999): 3–33.

[3] "A Definition of Irreversible Coma," *Journal of the American Medical Association* 205 (6) (1968): 337–40.

[4] While this language was commonly used in stating the whole-brain standard, it soon became obvious that there were sometimes "trivial" residual brain functions in individuals considered dead by the standard's proponents. As we will see, some critics have pressed this point in arguing that the standard is either incoherent or clinically inadequate.

of this standard, preferring to leave its application in the hands of the "conscientious physician."

Naturally, the committee did not work in a cultural and scientific vacuum. A 1959 French report had long before described *coma depasse*, a syndrome characterized by loss of all reflexes and brain activity in patients whose cardiopulmonary function could be sustained despite an irreversible loss of consciousness.[5] And several symposia and conferences that convened in the second half of the 1960s reinforced the notion that equating brain death with death, or recognizing brain death as one manifestation (along with cessation of cardiopulmonary function) of death, could help to resolve important practical problems.[6] Moreover, by the time the Ad Hoc Committee was formed, clinical practice sometimes proceeded on the implicit acceptance of brain death.

During the 1970s several states adopted statutes that recognized brain death either as the sole standard for determining death or as one standard alongside the traditional cardiopulmonary standard.[7] In states lacking such statutes, legal doctrine framed by judicial decisions embraced cardiopulmonary criteria. Lack of uniformity among states spurred the creation of model statutes by scholars and professional organizations such as the American Bar Association (ABA) and the American Medical Association (AMA). But the most influential model statute was the Uniform Determination of Death Act (UDDA), presented in a 1981 report, *Defining Death*, by the President's Commission for the Study of Ethical Problems in Medicine and Biomedical and Behavioral Research.[8]

Unlike the Ad Hoc Committee, the President's Commission recommended a uniform legal standard and provided a philosophico-conceptual rationale for it. Rather than *replacing* the widely recognized cardiopulmonary standard with a whole-brain criterion, a move that would have been more revisionary of existing law and clinical practice, the committee recommended a disjunctive standard that included both traditional and whole-brain criteria. Thus, "an individual who has sustained either (1) irreversible cessation of circulatory and respiratory functions, or

---

[5] P. Mollaret and M. Goulon, "Le Coma Depasse," *Revue Neurologique* 101 (1) (1959): 3–15.

[6] See Pernick, "Brain Death in a Cultural Context," p. 8.

[7] This paragraph has benefitted from Alexander Morgan Capron, "The Bifurcated Legal Standard for Determining Death," in Younger et al., *The Definition of Death*, pp. 120–5.

[8] *Defining Death: Medical, Legal and Ethical Issues in the Determination of Death* (Washington, DC: Government Printing Office, 1981). By the time of this publication, the National Conference of Commissioners on Uniform State Laws, the ABA, and the AMA had all adopted the UDDA.

(2) irreversible cessation of all functions of the entire brain, including the brain stem, is dead. A determination of death must be made in accordance with accepted medical standards."[9] The first sentence states that one is dead if *either* (1) or (2) applies. The second sentence wisely leaves open the clinical tests to be used in determining death, deferring to evolving medical expertise. At the time of the report, there was a working consensus that coma, apnea (absence of spontaneous efforts to breathe), and loss of brainstem reflexes were to be tested for.[10] The UDDA has been adopted by most American states, and all states now legally recognize the whole-brain standard, either disjunctively or exclusively – if not by statute, then by court decision.[11]

While the carefully crafted, minimally revisionary UDDA was ripe for widespread acceptance, the President's Commission enhanced this prospect with an ingenious philosophical rationale for the disjunctive standard. The rationale provides a *general definition or conception* of death. By contrast, *standards* or *criteria*, such as the traditional and whole-brain standards, are somewhat more specific and useful for legal purposes. According to the commission, death should be conceptualized as *permanent loss of integrated functioning of the organism as a whole*, a definition that applies not only to humans but to other animals and to plants as well.[12] Although a definition would seem necessary to make sense of a proposed standard of death, providing its conceptual rationale, the whole-brain standard enjoyed considerable support before the integrated-functioning definition was advanced by the commission and its clinical consultants, who published a seminal article in the same year as the report.[13]

While one might wonder whether it is anything more than a post hoc rationalization for a clinically convenient standard, the definition has

---

[9] Ibid., p. 119.

[10] Cranford, "Criteria for Death," pp. 530–1.

[11] Robert Olick, "Brain Death, Religious Freedom, and Public Policy: New Jersey's Landmark Legislative Initiative," *Kennedy Institute of Ethics Journal* 1 (4) (1991), pp. 275–6. New Jersey and New York, however, have instituted exemptions from whole-brain criteria for those who favor the traditional criterion (alone) on religious grounds – providing some informal evidence that the UDDA's standard marked a significant departure from the traditional understanding of death.

[12] President's Commission, *Defining Death*, pp. 32–3.

[13] James Bernat, Charles Culver, and Bernard Gert, "On the Definition and Criterion of Death," *Annals of Internal Medicine* 94 (1981): 389–94. The President's Commission and their clinical consultants were not the first to advance the organismic definition. Lawrence Becker, for one, offered essentially the same definition several years earlier ("Human Being: The Boundaries of the Concept," *Philosophy and Public Affairs* 4 [1975], p. 353).

undeniable merits. After all, the concept of death applies to all organisms. So, definitions specifying features such as consciousness that do not characterize all life forms do not seem to capture the concept. (Of course, one might argue that we should reconceptualize or reconstruct human death in a way that departs from this original biological understanding, and we will consider several attempts to do so.) Further, what occurs when a flower, tree, bug, turtle, or monkey dies is that *the organism's overall bodily system stops functioning as an integrated unit.* Articulating this conceptual common ground among accepted instances of death among the wide variety of lifeforms answers the objection that death might not be a unitary concept, susceptible to a single definition. The best way to prove that there is a single definition is to provide one that captures what is common to all instances of death – and the commission's definition does this with considerable plausibility.

Much more contentious is the commission's claim that, in offering a disjunctive standard that recognizes brain death, it wasn't redefining death but rather updating our understanding of how to determine or diagnose death. According to the commission, we have *always* understood human death as the breakdown of a system integrating cardiopulmonary and neurological functioning, rather than simply the discontinuation of cardiopulmonary function. Thus, the proposed disjunctive standard acknowledges the neurological window on human death, whereas the traditional standard acknowledges the cardiopulmonary window alone:

> Three organs – the heart, lungs, and brain – assume special significance... because their interrelationship is very close and the irreversible cessation of any one very quickly stops the other two and consequently halts the integrated functioning of the organism as a whole. Because they were easily measured, circulation and respiration were traditionally the basic "vital signs." But breathing and heartbeat are not life itself. They are simply used as signs – as one window for viewing a deeper and more complex reality: a triangle of interrelated systems with the brain at its apex. [T]he traditional means of diagnosing death actually detected an irreversible cessation of integrated functioning among the interdependent bodily systems. When artificial means of support mask this loss of integration as measured by the old methods, brain-oriented criteria and tests provide a new window on the same phenomenon.[14]

So, on this view, when a respirator is necessary for a patient to breathe and circulate blood (due to brain death), the technologies present a false appearance of life, hiding the fact that the organism as a whole

---

[14] President's Commission, *Defining Death*, p. 33.

lacks integrated functioning. Later we will consider challenges to this claim.

For now, note the contentiousness of these words from the first sentence: "the irreversible cessation of any one very quickly stops the other two...." This implies that brain death very quickly stops circulation and respiration. But many brain-dead patients on respirators have apparently breathed and maintained circulation for months. The President's Commission's position rests on the claim that such patients are not *really* breathing and maintaining circulation. What is occurring in the body only appears to be respiration and circulation: "This is 'breathing' and 'circulation' only in an analogous sense."[15] On its face, this is a remarkable claim.

A charitable interpretation of the commission's intent is that, although breathing and circulation occur in the body, the important point is the lack of integrated functioning: The brainstem is no longer integrating or regulating the other vital functions, so the organism as a whole no longer functions as an integrated unit. In this sense, according to the argument, it is the respirator – not the organism – that is responsible for the occurrence of respiration and circulation. This reasoning enjoys at least some initial plausibility as well as apparent compatibility with an organismic conception of death.

Less defensible, I suggest, is the claim that the disjunctive standard simply articulates what tradition always regarded as death while incorporating a more modern window on the phenomenon. One can imagine traditionalists saying of the brain-dead patient whose cardiopulmonary function is powered by a respirator, "I don't care *how* breathing and circulation are maintained, just that they *are* maintained."[16] The traditional standard, arguably, would not count such a patient as dead. On the other hand, the traditional and whole-brain standards – as well as the disjunctive standard – seem more or less compatible with the vague definition of death as loss of integrated functioning of the organism as a whole.[17] So in the technical sense of *definition*, which is contrasted with *standard*, the commission offered no new definition. Yet, in apparently

---

[15] Ibid., p. 37.

[16] Pernick argues that "the only section of the report to discuss history at all distorted the past...in words written by Beecher before the formation of the committee, over the objections of the committee's only professional historian" ("Brain Death in a Cultural Context," p. 23).

[17] Elsewhere I argued that the vague organismic definition did not determine a winner among these competing standards ("Persons, Organisms, and Death: A Philosophical

claiming that its position did not depart in any significant way from the traditional understanding of death, the commission stepped onto very questionable ground. Nevertheless, its model statute, the UDDA, gained widespread legal acceptance, and *Defining Death* proved to be enormously influential.

The movement in recent decades toward acceptance of neurological criteria for determining death has been international, not confined to the United States, although countries display interesting variations in their laws and practices. The United Kingdom, for example, embraces the *brainstem standard*: death as the irreversible cessation of brainstem function. Thus, parts of an individual's cerebrum might continue to function, though the individual qualifies as dead. But brainstem functions include not only primitive reflexes and spontaneous breathing, but also the operation of the ascending reticular activating system – or *reticular formation* – a sort of on/off switch that makes consciousness possible without affecting its contents. That means that brainstem death entails permanent loss of consciousness. Thus, while whole-brain death and brainstem death differ neurologically, they do not seem to differ in terms of the functions that we most prize, such as spontaneous breathing and consciousness. Indeed, the definition of human death offered by Chris Pallis, the original proponent of the brainstem standard, is *irreversible loss of the capacity for consciousness and of the capacity to breathe.*[18] Outside the English-speaking world, many or most nations, and virtually all First World countries, have adopted either brainstem or whole-brain criteria for the determination of death.[19]

Thus, the *brain death movement* – using the term broadly to include the brainstem death variation – has achieved a remarkable consensus among First World nations and, in the United States, among the various states. But the consensus, while broad, has not been very deep and is therefore somewhat unstable. It is challenged by both traditionalists who accept only the cardiopulmonary standard and by proponents of a more radical

Critique of the Higher-Brain Approach," *Southern Journal of Philosophy* 37 [1999], pp. 433–4).

[18] See Chris Pallis, *The ABC of Brain Death* (London: British Medical Journal Publishers, 1983) and "On the Brainstem Criterion of Death," in Younger et al., *The Definition of Death*: 93–100.

[19] A recent study found that practice guidelines for determining brain death were present in seventy out of eighty countries surveyed (Eelco Wijdicks, "Brain Death Worldwide: Accepted Fact But No Global Consensus on Diagnostic Criteria," *Neurology* 58 [January 2002]: 20–5).

standard. Within both groups there are some who argue that brain death is an incoherent concept.[20]

Among the traditionalists in the United States we may include many Orthodox Jews and conservative Christians who have argued that the notion of brain death as (a type of) death violates our time-honored understanding of the phenomenon. Both New Jersey and New York have created legal space for these individuals to exempt themselves from whole-brain criteria. Among scholars, several recent publications have contended, independently of appeals to religion or even tradition per se, that whole-brain death is insufficient for human death: Sometimes brain-dead individuals on life supports maintain too much integrated functioning to count as dead (a thesis we will later examine more closely).[21] And, in several cultures that have legally accepted brain death, such as Japan and Germany, there remains deep ambivalence about the idea that brain death entails human death.[22]

On the other side of the debate are commentators who believe that the movement away from tradition toward neurological criteria has not gone far enough. They advocate the *higher-brain standard*: human death as *the irreversible cessation of the capacity for consciousness*.[23] For advocates of the higher-brain standard, what's important about the brain is its making psychological life possible, so irretrievable loss of the latter constitutes

---

[20] In an issue of *The Journal of Medicine and Philosophy* entitled "Revisiting Brain Death" (26 [5] [October 2001]), every article questions the coherence of this concept.

[21] See, e.g., D. Alan Shewmon, "The Brain and Somatic Integration: Insights into the Standard Biological Rationale for Equating 'Brain Death' with Death," *Journal of Medicine and Philosophy* 26 (2001): 457–78; and Michael Potts, "A Requiem for Whole Brain Death: A Response to D. Alan Shewmon's 'The Brain and Somatic Integration'," *Journal of Medicine and Philosophy* 26 (2001): 479–92.

[22] See Margaret Lock, "The Problem of Brain Death: Japanese Disputes about Bodies and Modernity," in Youngner et al., *The Definition of Death*: 239–56; and Bettina Schoena-Seifert, "Defining Death in Germany: Brain Death and Its Discontents," in Youngner et al., *The Definition of Death*: 257–71.

[23] See, e.g., Robert Veatch, "The Whole-Brain-Oriented Concept of Death: An Outmoded Philosophical Formulation," *Journal of Thanatology* 3 (1975): 13–30; Tristam Engelhardt, "Defining Death: A Philosophical Problem for Medicine and Law," *Annual Review of Respiratory Disease* 112 (1975): 312–24; Michael Green and Daniel Wikler, "Brain Death and Personal Identity," *Philosophy and Public Affairs* 9 (1980): 105–33; Karen Gervais, *Redefining Death* (New Haven, CT: Yale University Press, 1986), ch. 6; Edward Bartlett and Stuart Youngner, "Human Death and the Destruction of the Neocortex," in Richard Zaner (ed.), *Death: Beyond Whole-Brain Criteria* (Dordrecht, the Netherlands: Kluwer, 1988): 199–216; Roland Puccetti, "Does Anyone Survive Neocortical Death?" in Zaner, *Death*: 75–90; Richard Zaner, "Brains and Persons: A Critique of Veatch's View," in Zaner, *Death*: 187–97; and Ben Rich, "Postmodern Personhood: A Matter of Consciousness," *Bioethics* 11 (1997): 206–16.

the death of a human being. Hence, a *psychological* definition of human death: irretrievable loss of personhood or psychological life. The psychological definition emphasizes our nature as persons or conscious subjects, whereas the organismic definition stresses our biological nature. According to the higher-brain approach, and contrary to current legal standards and medical practice, individuals in a PVS or permanent coma are dead despite the continuation of at least some brainstem functions.[24]

While popular among philosophers, the higher-brain standard has never been adopted by any country or state. There appear to be two major reasons for this. First, the higher-brain standard implies that a spontaneously breathing patient in PVS or permanent coma is dead. Many people find this judgment highly counterintuitive, and the apparent practical implication that it would be appropriate to bury or cremate such an individual, repugnant. (Whether this really is an implication of the higher-brain standard is debatable. One might argue that we should not bury or cremate an individual in PVS or permanent coma out of deference to common esthetic sensibilities, not because the individual remains alive.) Second, the higher-brain standard is much more obviously a departure from tradition than is the disjunctive approach, which retains the cardiopulmonary standard and remains within a broadly organismic understanding of life and death; the higher-brain standard is also more antitraditional than a pure whole-brain or brainstem standard. For this reason, proposing the higher-brain standard arouses suspicions that transplant surgeons, cost-conscious hospital executives, and other interested parties would prematurely declare death to pursue their agendas. While any departure from tradition risks arousing such suspicions – I have suggested that even the disjunctive UDDA approach departs somewhat from tradition – the latter approach appeared moderate enough to minimize such concerns while capitalizing on another: that physicians using only the cardiopulmonary standard would indefinitely maintain brain-dead patients with invasive medical technology.[25]

Many commentators believe that, despite these obstacles to acceptance, the higher-brain standard is the most coherent and plausible

---

[24] Much of the literature on death conflates permanent coma and PVS, although PVS patients have open eyes and in other ways (misleadingly) appear wakeful. I will reserve the term *PVS* for those who are in *permanent* vegetative states; many clinicians consider vegetative states *persistent* whenever they endure longer than one month (even if they are reversible). For a helpful discussion of various states of permanent unconsciousness, see Cranford, "Criteria for Death," pp. 531–3.

[25] See Pernick, "Brain Death in a Cultural Context," p. 4.

account of human death. Courtney Campbell, for example, argues that the difficulties of any organismic approach leave open "a social move to higher brain death as a plausible policy option, and one that can meet tests of philosophical coherence."[26] I believe that this confidence is badly misplaced and that the higher-brain approach is not only difficult to accept psychologically and politically, but conceptually dubious as well. Unfortunately, most critiques of the higher-brain approach have been philosophically superficial.[27]

Much of the remainder of this chapter is devoted to countering the strongest philosophical defenses of the higher-brain standard, with special attention to the role of personal identity theory. I counter in turn appeals to person essentialism, appeals to the psychological view of numerical identity, and appeals to narrative identity. I then address arguments that rest on the claim that death is a moral concept before exploring which standard the organismic approach most strongly supports. Finally, I conclude with comments about policy options.

CRITIQUE OF APPEALS TO OUR ESSENCE IN DEFENSE OF THE HIGHER-BRAIN STANDARD

Many proponents of the higher-brain standard of death reason in a way that may be reconstructed as follows:

(1) For humans, permanent loss of the capacity for consciousness is loss of personhood.
(2) For humans, loss of personhood is death.
Ergo
(3) For humans, permanent loss of the capacity for consciousness is death.

The argument concludes by asserting the higher-brain standard. What is most interesting about the argument, however, are its premises. Is it fair of me to claim that many champions of the higher-brain view assert these premises?

Consider the following statements by philosophers and bioethicists. Lynn Rudder Baker: "When the organism can no longer support a

---

[26] "A No-Brainer: Criticisms of Brain-Based Standards of Death," *Journal of Medicine and Philosophy* 26 (2001), p. 547.
[27] See, e.g., Charles Culver and Bernard Gert, *Philosophy in Medicine* (New York: Oxford University Press, 1982), ch. 10. I attempt a more substantial critique in "Persons, Organisms, and Death."

first-person perspective, then it no longer constitutes a person. And if something ceases to be a person, it ceases to be – even if the human organism that constituted the person continues to exist."[28] Richard Zaner: "[T]he permanent loss of personhood *is* death in the most significant sense."[29] Ben Rich: "Robert Veatch [correctly] suggests that death should mean 'a complete change in the status of a living entity characterized by the irretrievable loss of those characteristics that are essentially significant to it'.... What is unique about human beings is their capacity for personhood, for living the self-conscious life of a person."[30] Edward Bartlett and Stuart Youngner: "Loss of personhood cannot be understood as the loss of an attribute, which could be followed by death. Once it has been lost, the subsequent death is not that of a human being; instead, it is the death of a thing."[31] Tristam Engelhardt: "Such bodies, since they are no longer the embodiment of persons or even a mind, would count, at least in general secular terms, as biologically living corpses."[32] Robert Veatch: "[T]he human is essentially the integration of the mind and body and the existence of one without the other is not sufficient to constitute a living human being."[33] Karen Gervais: "Unless our concept of human death reflects what is essentially significant to the nature of the human being – conscious awareness – it fails to provide a community with an effective moral divide between the living and the dead."[34]

The statements that mention personhood suggest that we human persons – that is, we human individuals who are (now) persons – are essentially persons. Those using the term *human being* imply that we are essentially human beings. But the latter term is clearly used not in a purely biological sense extending to all *Homo sapiens*, but at least partly in a psychological sense: human beings as essentially conscious or minded beings (and perhaps also essentially embodied). This psychological sense of *human being* is similar to either (1) the sense of *person* introduced in Chapter 1 or (2) the sense of *mind* or *minded being* discussed in Chapter 2, which refer to beings with the capacity for consciousness. So each of the

---

[28] *Persons and Bodies* (Cambridge: Cambridge University Press, 2000), p. 5.
[29] "Introduction," in Zaner, *Death*, p. 7.
[30] "Postmodern Personhood," p. 121.
[31] "Human Death and the Destruction of the Neocortex," pp. 210–11.
[32] *The Foundations of Bioethics*, 2nd ed. (New York: Oxford University Press, 1996), p. 248.
[33] "The Impending Collapse of the Whole-Brain Definition of Death," *Hastings Center Report* 23 (July–August 1993), p. 21.
[34] "Definition and Determination of Death: Philosophical and Theological Perspectives," in Reich, *Encyclopedia of Bioethics*, p. 547.

authors just quoted implies either that we human persons are essentially persons or that we are essentially minds. The most natural reading of this essentialist claim is the metaphysical thesis of person (or mind) essentialism, as discussed in Chapter 2, which states necessary conditions for our continued existence or numerical identity. (Later, we will discuss a value-based sort of essentialism.) Let us begin with the reading that attributes person essentialism, beginning with a closer examination of the argument's premises.

Premise (2) asserts person essentialism. Premise (1) concerns the nature of personhood, stating that the capacity for consciousness is necessary for being a person – consistent with Chapter 1's conception of personhood. But when we examine the premise more closely, problems appear. At the risk of sounding like Bill Clinton, we must ask about the meaning of *is* in "permanent loss of the capacity for consciousness *is* loss of personhood." Does it mean "equals" or merely "entails"? Clearly, on our conception of personhood, permanent loss of the capacity for consciousness *suffices* for loss of personhood – a claim that implies the uncontroversial thesis that this capacity is necessary for personhood. But is permanent loss of the capacity for consciousness also *necessary* for loss of personhood, implying that this capacity suffices for personhood? Either way, those who reason along the lines of the preceding syllogism will be impaled on the horns of a dilemma.

Suppose that *is* means "equals," so that permanent loss of the capacity for consciousness is both necessary and sufficient for loss of personhood. This implies that the capacity for consciousness is not only necessary but also sufficient for being a person. In that case, all sentient animals, who by definition have the capacity for consciousness, are persons. Given available empirical evidence, this would suggest that most or all vertebrates are persons[35] – an assertion so implausible that it suggests that this conception of personhood cannot be correct.

Could one defend the "necessary and sufficient" reading of premise (1) by stressing its first two words, "For humans"? Maybe the capacity for consciousness is necessary and sufficient for personhood in the case of humans but not other animals. But this move has one of two problems, depending on how we unpack it. If it asserts different conditions for personhood for different sorts of beings, it invites the charge of incoherence

---

[35] For philosophical arguments and citations of empirical literature supporting the claim that most or all vertebrates are sentient, see my *Taking Animals Seriously: Mental Life and Moral Status* (Cambridge: Cambridge University Press, 1996), ch. 5.

for failing to capture a single conception of personhood. If, on the other hand, it adds to the capacity for consciousness another condition for personhood – membership in our species – then it violates the reasonable expectation that an analysis of personhood be species-neutral, stating conditions that in principle multiple kinds of creatures (say, from other planets) or a Creator could satisfy. So stressing "For humans" will not help.

The more plausible reading of premise (1) takes *is* in the sense of "entails" or "is sufficient for": Permanent loss of the capacity for consciousness is sufficient (but not also necessary) for loss of personhood. This implies that the capacity for consciousness is necessary but not sufficient for personhood, a claim that fits with the conception of personhood defended earlier: persons as beings with the capacity for *sufficiently complex* forms of consciousness (such as planning and fairly developed self-awareness). But, while this reading of premise (1) makes sense of personhood, it makes nonsense of premise (2).

How so? The thesis that personhood requires the capacity for sufficiently complex forms of consciousness implies that a being might have the capacity only for *insufficiently complex* forms of consciousness and thereby fail to be a person. Indeed, this is the case for the vast majority of sentient nonhuman animals. But now consider a relevantly similar human being – say, a woman in such an advanced stage of Alzheimer's disease that she fails to meet conditions of personhood. Premise (2) says that loss of personhood, for humans, *is* death. It doesn't matter whether *is* here means "equals" (is both necessary and sufficient for) or merely "entails" (is sufficient for); either way, loss of personhood suffices for death. *This apparently implies that the sentient nonperson with advanced Alzheimer's disease is dead.* Yet she may still be able to shuffle around, mumble, perform very simple intentional actions, and the like. To suggest that this human being is dead is completely implausible and, given the natural inference that she lacks moral status, morally outrageous.[36]

---

[36] One won't find this inference natural if one holds that deceased individuals have some degree of moral status, explaining our norms about not desecrating corpses or maliciously slandering the deceased. But, even if deceased individuals have some moral status, they have considerably less than living persons – so equating the moral status of a severely demented individual with that of the deceased would still attribute very little to the former, much less than we are inclined to attribute. Moreover, it's implausible to claim that the deceased have moral status. As argued in Chapter 2, we go out of existence when we die. The corpse, or pile of ashes following cremation, is not something

However, Baker's preceding statement suggests a somewhat more congenial possibility: The sentient nonperson with Alzheimer's disease remains alive as a human organism, although the person who was constituted by this organism has gone out of existence or died. In Chapter 2, I argued against Baker's thesis that human organisms constitute persons, her thesis that our identity consists in a continuing first-person perspective, and person essentialism generally. Any view that buys the plausible claim that the sentient nonperson with Alzheimer's disease is alive at the price of person essentialism has paid too much. We are not essentially persons, and that is why we can anticipate (to some extent) and fear becoming severely demented.

Another problem with premise (2) is that it doesn't apply to human beings who have never been persons. Since such human beings cannot lose personhood, under what conditions do they die? Consider sentient humans who have never been persons, such as the most severely retarded children. They are unquestionably alive, but premise (2) provides no criteria for their death; someone who holds the present view must argue that these children are of a radically different kind of being than you and I instantiate, with a different essence and identity conditions. Consider anencephalics, who never will have the capacity for consciousness due to the absence of a functioning cerebrum. Some who accept premise (2) may assert that anencephalic fetuses and infants are never alive. Considering their brainstem function and unassisted breathing, many will find this claim counterintuitive. Another option is to maintain that anencephalics, although alive, are of a different kind from human persons.[37] In any case, the problem that premise (2) does not apply to humans who have never been persons is a problem of incompleteness. It can be remedied either by arguing that these individuals are never alive or by providing an account of death for them and justifying the assertion of different criteria for their death and for ours. Such a justification would appeal to the essence of human persons, which, according to this viewpoint, the other humans lack. Assuming that the essence is personhood itself, as in person essentialism, this development of the view in question will not prove very attractive.

that one of us can become. Lacking even a *history* of sentience, what possible claim can this material have to moral status? The intuitions that may seem to motivate the claim that the dead have moral status – such as prohibitions against desecrating corpses and speaking maliciously of the dead – are better explained in terms of norms for expressing respect for the person who was. Chapter 5 will revisit this topic.

[37] Green and Wikler take this approach ("Brain Death and Personal Identity," p. 128).

A less problematic position embraces mind essentialism, the thesis that we are essentially beings with the capacity for consciousness, a capacity both necessary and sufficient for the existence of beings of our kind. This is Jeff McMahan's view and apparently that of the authors quoted earlier who used the term *human being* rather than *person*.

Let us revise the preceding syllogism so that it assumes mind essentialism, not person essentialism, and otherwise reads as plausibly as possible:

(1) For sentient beings, permanent loss of the capacity for consciousness (sentience) is loss of a mind.

(2) For sentient beings, loss of a mind is death.

Ergo

(3) For sentient beings, permanent loss of the capacity for consciousness (sentience) is death.

I change the scope of the argument from humans to sentient beings because I can imagine no good reason why some minded or sentient beings, such as (normal) human beings, would be essentially minded, while others would not be. Our revised syllogism expresses McMahan's view about the death of minded beings. Interestingly, on this view, there are two deaths associated with each of us: We die on permanently losing the capacity for consciousness; "our organisms" die when they suffer a terminal cardiac arrest.[38] The PVS patient, then, is merely a human organism, not one of us. One advantage of this view is that the spontaneously breathing PVS patient counts as alive, consistent with common intuitions, as does the anencephalic.

Another version of mind essentialism holds that there is only one death associated with each of us: You die when you permanently lose the capacity for consciousness, and if a spontaneously breathing human organism remains, it is dead. Roland Puccetti holds this view.[39] I find it hard to believe that a human organism could be dead while spontaneously breathing, circulating blood, metabolizing, excreting wastes, and exhibiting a full range of brainstem reflexes. Our shared concepts of life and death, after all, are biological – and from a biological standpoint, the fact that this individual is permanently unconscious seems irrelevant, while the

[38] *The Ethics of Killing: Problems at the Margins of Life* (Oxford: Oxford University Press, 2002), pp. 423–6.

[39] He puts it strikingly: "corpses are really of two kinds: the vast majority that cannot breathe unaided, and a small minority that nevertheless can do this" ("Does Anyone Survive Neocortical Death?" p. 85).

well-integrated forms of somatic functioning seem relevant. Now, the higher-brain theorist might intend to *reconstruct* the concept of death in the case of humans (or some broader class of beings such as sentient animals) in a way that departs from the shared biological concept of life. But the burden of persuasion is on such theorists, and I will argue that none of them successfully shoulders this burden. For now, I will assume that the two-deaths version of mind essentialism is more promising for allowing that nonsentient yet spontaneously breathing human organisms are alive. And since it is clearly more promising than any view that assumes person essentialism, I assume it is the most viable view among those that understand our identity and essence in psychological terms.

But, having argued that mind essentialism is less plausible in the end than the biological view, according to which we are essentially human animals, I must reject mind essentialism's implicit endorsement of the higher-brain standard. While mind essentialism avoids some of the difficulties associated with person essentialism (see Chapter 2), it retains two major difficulties.

First, it implies that, strictly speaking, we are not animals. Having persistence conditions that can and sometimes do depart from those of human animals, we are not identical to them. While I cannot bring myself to believe this, some people apparently can.

The most devastating problem of mind essentialism, a problem common to all psychological views, is its apparent inability to explain plausibly the person–human organism relationship. So far, no defense of the claim that the human organism *constitutes* the person has been persuasive. And McMahan's suggestion that the person or mind *is part of* the organism, we found, led to problems no matter how we developed it: either substance dualism, or the thesis that a person is simply a set of properties, or the claim that each of us is a brain. The last possibility seemed the least unpalatable, yet it encountered two major problems. First, either we continue to exist as dead, nonfunctioning brains, contradicting the claim that we are essentially beings with the capacity for consciousness, or we are *brains functioning in certain characteristic ways* – yet the thesis that a functioning brain is a distinct substance from a brain *simpliciter* (which can function or not) is hardly plausible. Second, if we are brains, then we have many properties we don't normally believe ourselves to have, such as weighing less than ten pounds and being gray and convoluted even as young people. I therefore conclude that appeals to neither person essentialism nor mind essentialism will vindicate the higher-brain standard.

CRITIQUE OF APPEALS TO PERSONAL IDENTITY IN DEFENSE OF THE HIGHER-BRAIN STANDARD[40]

Another strategy for supporting the higher-brain standard is to appeal to personal identity (in the numerical sense) without making contentious assumptions about our essence. Considering the claim advanced in Chapter 2 that numerical identity and essence are logically connected, this strategy may seem dubious. But, since its development by Michael Green and Daniel Wikler, it has frequently been cited as a strategy distinct from appeals to our essence.[41] Moreover, Green and Wikler appeal to our essence *as individuals* – that is, embrace a type of *individual essentialism* – rather than appealing to our essence as *members of some kind* (such as persons or human animals); and my claim about the essence–identity link implicitly addressed *kind* essentialism. So our preceding arguments are insufficient to undermine this approach.

Green and Wikler attempt to defend the higher-brain approach without invoking person essentialism or an account of the nature of persons. Instead, they appeal to a version of the psychological approach to identity.[42] Adducing familiar arguments about the intelligibility of body switching, and updating the thought experiments with reality constraints grounded in contemporary neurology, the authors assert that personal identity is closely tied to brain identity.[43] If a brain is destroyed, they continue, clearly the associated person's history has ended. Similarly, a brain-dead body lacks the identity of the previously associated person.[44]

---

[40] Much of this section draws from a discussion in my "Persons, Organisms, and Death," pp. 425–9.

[41] For example, in discussing arguments for the higher-brain standard, the President's Commission distinguishes appeals to personhood (which assume person essentialism) and appeals to personal identity (*Defining Death*, pp. 38–9).

[42] They align themselves with the work of John Perry. Surprisingly, they attribute to Perry the thesis that "two 'person-stages' are stages of the same person, just in case the later is a continuation of the earlier personality *and can remember what the earlier has done*" ("Brain Death and Personal Identity," p. 121, my emphasis). As Perry himself notes, this claim, which Locke apparently made, absurdly implies that one must be able to remember everything one has ever done (Perry, "The Problem of Personal Identity," in Perry [ed.], *Personal Identity* [Berkeley: University of California Press, 1975], p. 15). Most psychological theorists have modified Locke's approach to avoid this difficulty.

[43] "Brain Death and Personal Identity," p. 124.

[44] Ibid., p. 126. They use the term *brain death* more broadly than it is used now, referring to both whole-brain and higher-brain death. But they leave no doubt that they defend the latter standard.

The death of a person, they conclude, consists in the permanent cessation of higher-brain function.[45]

My reply, in a nutshell, is that Green and Wikler cannot avoid assuming a type of kind essentialism – either person essentialism or mind essentialism – and therefore cannot avoid associated difficulties. Consider this summary passage, with sentences numbered for easy reference:

[1] [T]he continued possession of certain psychological properties by means of a certain causal process [that is, psychological continuity] is an essential requirement for any given entity to be identical with the individual who is Jones. [2] Thus, we can afford to remain uncommitted on whether persons are essentially beings with psychological properties and on whether Jones is essentially a person. [3] We demonstrate instead that Jones, whatever kind of entity he is, is essentially an entity with psychological properties. [4] Thus, when brain death strips the patient's body of all its psychological traits, Jones ceases to exist.

[5] What is required to show that Jones is an individual who essentially possesses certain psychological capacities is an adequate account of personal identity.[46]

Concerning [1], which psychological properties are necessary for Jones to continue to exist? The higher-brain standard suggests that *any* psychological or conscious life will suffice. If the authors imply, despite themselves, that we are essentially persons and that the latter are essentially psychological beings, they will have trouble maintaining that any conscious life suffices. But if conscious life is insufficient, person-typical properties such as the ability to plan and self-awareness also being necessary, then the authors inherit the various problems associated with person essentialism – such as the implication that Jones was never born (see Chapter 2).

That brings us to a more fundamental problem. Statement [1] refers to "an essential requirement for any given entity to be identical with the individual who is Jones." Identical *by what criterion*? Like any other entity, Jones instantiates more than one kind of thing. From the standpoint of common sense, he is a person, a sentient being, a member of *Homo sapiens*, a primate, an organism, and so on. Among the many kinds of thing that Jones is, the authors must privilege one of them to make their claim about Jones's identity. Even if one rejects the commonsense claim that Jones instantiates all the categories mentioned – holding, for example, that Jones, who is a person and therefore a sentient being, is not,

45 Ibid., p. 127.
46 Ibid., p. 121.

strictly speaking, a member of *Homo* sapiens, a primate, or an organism – this claim would also require an assertion of kind essentialism. Either way, *the authors' individual essentialism logically depends on some type of kind essentialism.*

Clearly, Green and Wikler do not favor a biological criterion for Jones's identity.[47] Indeed, they imply that Jones is, strictly speaking, distinct from the organism associated with him. Permanent loss of consciousness doesn't guarantee the death of an organism, so Jones's loss of identity and consequent "death" must be understood by reference to some other criterion corresponding to some kind. So the authors must assert both that Jones is essentially a member of some kind and (given [1]) that an essential characteristic of members of that kind is the capacity for consciousness.

So what kind of being is Jones essentially? One such that its members cease to exist on losing their psychological life. And presumably it's one that you and I instantiate along with Jones, since we are told virtually nothing about him in particular. Sentence [5] may furnish the answer inasmuch as theorists often take a theory of *personal* identity to state the persistence conditions of beings who are assumed to be essentially persons.[48] Not surprisingly, the language of "persons" occurs throughout their discussion.[49] The following statement is illuminating, because it reveals that the relevant kind is one that Jones instantiates while anencephalics do not: "... the identity criteria for the anencephalic, never-to-be conscious infant do not involve causal substrates for higher level psychological continuity. The conditions for life and existence will be those of human bodies rather than those of *persons*."[50]

Thus, on the most natural reading of this account, Jones is essentially a person, and persons are essentially beings with psychological capacities. Thus [2], which disavows both forms of essentialism – kind essentialism and a view about the nature of persons – is unsustainable. Sentence [3] – which holds that Jones, whatever kind of thing he is, essentially has

---

[47] See, e.g., ibid., p. 118, n. 24.

[48] This is not strictly necessary, however, as explained in Chapter 1. One might say that a personal identity theory provides persistence conditions of beings who are *ever* persons, whatever kind of being they are essentially. Thus, Eric Olson, who holds that we are essentially human animals, refers to his theory as one of personal identity. Perhaps it is preferable to refer to *our identity* or *human identity*, phrases that have no tendency to suggest person essentialism.

[49] I dispute the authors' statement that their constant use of the term *person* is merely for convenience ("Brain Death and Personal Identity," p. 121).

[50] Ibid., p. 128 (emphasis added).

psychological properties – begs the question. If Jones is essentially a human animal, say, he is not essentially a being with psychological properties. To insist that he is requires privileging some psychological kind; again, it appears that the authors implicitly invoke personhood. Finally, [4], which asserts that the loss of psychological properties entails Jones's death, is highly contentious, effectively begging the question of Jones's essence.

Meanwhile, another way to interpret the Green–Wikler view rescues it from some of these difficulties. In view of the plurality of kinds Jones instantiates, the authors must embrace a form of kind essentialism. But the kind in question might be *minded beings* – beings with the capacity for consciousness. If Jones is essentially a minded being, then he is "essentially an entity with psychological properties" (as [3] states). Jones's permanent loss of the capacity for consciousness would indeed entail his death, as the higher-brain standard implies.

We have already seen that mind essentialism has some advantages over person essentialism. We have also seen that it has major difficulties, supporting the conclusion that the biological view of our identity and essence is more viable. Moreover, Green and Wikler do not even defend the kind essentialism they assume – whether person essentialism or mind essentialism – unsurprisingly, as they do not realize they are assuming any such account. So the authors have certainly not overturned the presumptive biological concept of death with a persuasive case for the higher-brain standard. The strategy of appealing to personal identity appears unpromising.

CAN APPEAL TO NARRATIVE IDENTITY SUPPORT THE HIGHER-BRAIN VIEW?

The strategy reviewed in the previous section appealed to the numerical identity and, implicitly, the essence of human persons. Might an appeal to *narrative* identity prove more successful in supporting the higher-brain approach? Perhaps a few of the scholars quoted earlier who appeared to embrace person or mind essentialism really had in mind narrative identity – and a corresponding value-based essentialism.

I imagine an appeal to narrative identity running something like this:

Suppose the biological view is correct concerning our numerical identity. Then Chiang might lose the capacity for consciousness but continue to exist in PVS or permanent coma. Still, Chiang's self-narrative is now complete – there is no possibility of his adding to the story he had been telling himself about his own life. In this narrative sense, he no longer maintains his identity. So we should *regard* him as dead, even though he is biologically alive. In the same way, we should *regard* permanent loss of the capacity for consciousness as death in the case of all human persons, who are the sorts of beings who have self-narratives. For practical purposes, we should extend this standard to all human beings who exhibit signs of any mental life at all, thereby avoiding agonizing line-drawing regarding who is a person and who is merely sentient.

This appeal to narrative identity is attractive. Setting aside ontological issues, such as that of our essence, it defends the higher-brain standard on the basis of what's most important in our lives: conscious life, a precondition for the unfolding self-narratives characteristic of persons. The claim is not that the capacity for bare consciousness is *sufficient* for what's important in our lives, just that it is necessary – and represents the best place to draw a line for policy purposes while doing justice to what's important.

At bottom, then, the appeal to narrative identity is a value-based argument. More specifically, it concerns "what matters in survival" (as discussed in Chapters 2 and 3). And a value-based argument is the form of argument one would expect for a conclusion that *we should regard* permanent loss of the capacity for consciousness as death. (By contrast, we would expect arguments that such-and-such *is* death to be ontological or conceptual.) It is helpful to note the value-based nature of the present argument, in order to clarify the terms in which evaluation of the argument should be carried out: value terms. But let us sharpen the argument a bit more before evaluating it.

First, the appeal to narrative identity primarily concerns *prudential* value, what matters from a self-regarding standpoint: We have an enormous stake in continuing our self-narratives, other things equal, and little or no stake in continuing our existence when we have become permanently unconscious.[51] The capacity for consciousness is essential in the sense of *indispensable to us*. While those who are sympathetic to this reasoning may also adduce certain moral considerations – such as the desirability of not devouring hospital resources by caring for patients who, arguably, can no longer benefit – such considerations are merely complementary to the present argument. Second, the appeal to narrative identity need not claim that the capacity for consciousness underlies *everything* of

---

[51] Marya Schechtman argues more or less along these lines (*The Constitution of Selves* [Ithaca, NY: Cornell University Press, 1996], pp. 150–2).

prudential value, just that it underlies the overwhelmingly greater part of what we value prudentially.

Both points provoke questions, however. First, why should prudential value be the primary basis of what we regard as human death? Why should moral values be only complementary? If prudential and moral values conflict about which standard of death is preferable, what then? Suppose Chiang regarded consciousness as a precondition for what mattered in his life, a point that favors the higher-brain standard. But Chiang's family disagrees. They would like – after Chiang's horrible car accident, say, in which he entered PVS – several weeks during which they can visit Chiang in his hospital room, talking to him as he breathes unassisted (even though he can't really hear them), getting used to his condition on the road to death. Needing time for closure on Chiang's psychological demise, they value his continued biological life as a means to this end, so they prefer not to regard his permanent loss of the capacity for consciousness as his death. From a moral standpoint, arguably, their interest carries some significant weight. It is hardly self-evident that only the prudential values of the individual who might be considered dead should count in a value-based approach to determining death.

Nevertheless, there is reason to favor prudential values. Just as a competent adult should ordinarily be permitted to consent to or refuse medical treatments that are made available to her, on the basis of her own understanding of her best interests (or prudential value), a competent adult should be allowed to determine that permanent loss of the capacity for consciousness will count as death in her case on the basis of her self-regarding values. But why should a patient's choice about medical treatments or an individual's choice about what will constitute her death take into account only her *self-regarding* values, as opposed to her values more generally? After all, a competent adult may wish to incorporate moral values like the desirability of his family's opportunity to achieve closure while he is in PVS. His standpoint of "overall preferability" in this context might not be purely prudential. One might reply that, in this sort of case, the individual has transformed certain moral considerations into prudential ones, so that the family's interests become part of his interests. While I am skeptical of the proposition that *anything* a patient values can be incorporated into his prudential values, let's grant this proposition – at least for argument's sake. Effectively the same issue arises *within* the realm of prudential values, and resolution of the issue will not vindicate the higher-brain standard for everyone.

Basing the higher-brain standard on an appeal to prudential values confronts the difficulty of pluralism within prudential values. Some people prudentially value things that do not require the continuing capacity for consciousness. If a patient cares (prudentially, let's say) for his family's need for closure while he is in PVS, that would count against the higher-brain standard in his case. Or imagine a fundamentalist Christian or an Orthodox Jew who believes that (biological) life is inherently precious to its possessor, whether or not she can appreciate its value at a given time. This, too, would count against the higher-brain standard in this individual's case. The appeal to prudential value seems to favor *a pro-choice view about standards of death*, not the higher-brain standard in particular.

This point can be stated in terms of narrative identity. Imagine a patient whose narrative prominently features his family members and portrays the patient himself as fundamentally a member of this family, constituted (in a value-based, not metaphysical, sense) partly by his relations with family members. Given their need for his continuing, breathing presence in the hospital room, the most coherent and faithful conclusion of his self-narrative might require that he live for a few weeks in PVS. Granted, for this argument to work, the metaphor of a self-narrative must be flexible enough to allow someone other than the autobiographer to write the last chapter, or to allow the original author to complete his story in advance and to authorize others to bring the last chapter to fruition. But this interpretation of self-narrative seems reasonable. (Chapter 5 will more fully defend the claim that an agent can extend her narrative beyond the days when she has *narrative capacity*.) Meanwhile, the other patient we imagined, who is strongly pro-life, clearly sees his life as continuing until he meets some appropriate biological standard of death. His self-narrative anticipates the possibility of PVS or permanent coma, let's say, and declares that it may ultimately include a period of time in one of these states. Just because he can't self-consciously continue to write his self-story while unconscious is no more reason to think his story can't include this period, one might argue, than the fact that Gramma is senile is a reason to think her advance directive, which she completed while competent, cannot be authoritative while she is incompetent.

One might reply that to value biological existence without consciousness is *irrational* and therefore provides no support for pluralism regarding standards of death. From this standpoint, rational prudential evaluation singles out the higher-brain standard.

But the charge of irrationality is highly contentious. It assumes, in effect, the *experience requirement*: that only states of affairs that affect one's experience can affect one's well-being.[52] But many people, on reflection, think otherwise. They may believe, for example, that they are worse off if slandered even if they never hear of the malicious talk and none of its consequences enter into their experience. Some believe that the quality of one's life as a whole can be affected by posthumous states of affairs such as the murder of a loved one. Although I doubt the intelligibility of this thesis – and will argue against it in Chapter 5 – I find the following more modest claim quite intelligible: States of affairs that don't affect one's experience but connect importantly with one's values can affect one's interests or well-being *while one exists.* While existing is a precondition for being harmed or benefited, having one's consciousness affected is not necessary. Desire-based accounts of well-being standardly accept this principle, because what is desired may take place without one's awareness of its taking place and without affecting one's awareness. The desire can be satisfied, we might say, without the desirer *feeling* satisfied. If this is correct, it opens the door to the intelligibility of having prudential values that extend to a time when one is permanently unconscious (yet still alive). While I am inclined to accept the experience requirement in my own case, I am not aware of persuasive arguments that establish it as a general requirement of rational prudential evaluation.[53]

Thus, it appears that the appeal to prudential value does not vindicate the higher-brain approach as a universal standard. The everyday organismic concept of death, which is incompatible with this standard, remains standing; no argument in favor of reconstructing human death has overturned the presumption favoring the original concept of death. But I suggest that the appeal to narrative identity and prudential value does justify permitting conscientious exemptions to whatever standard is ultimately specified within the organismic approach. Just as we permit competent adults to consent to or refuse medical treatment on the basis of their values and preferences, we should allow competent adults to exempt themselves from the default legal standard in favor of some reasonable alternative, such as the higher-brain criterion. We will revisit this issue in this chapter's final section. For now, let us examine one further effort to establish the higher-brain standard for everyone.

[52] For a good discussion of this putative requirement, see James Griffin, *Well-Being: Its Meaning, Measurement, and Moral Importance* (Oxford: Clarendon, 1986), pp. 16–19.

[53] Raymond Martin cogently argues that the appropriate focus of "what matters in survival" is what does, in fact, matter to us rather than what *should* matter to us (where the latter is based on debatable requirements of rationality). See *Self-Concern: An Experiential Approach to What Matters in Survival* (Cambridge: Cambridge University Press, 1998), pp. 18–27.

MIGHT DEATH BE A MORAL CONCEPT THAT IS INDEPENDENT OF IDENTITY THEORY?[54]

Appeals to person essentialism, numerical identity, and narrative identity have failed to justify reconstructing human death in accordance with the higher-brain standard (except as an individual exemption from a default standard). Let us examine one final reconstruction effort: the argument that the definition of death is a moral issue, and that confronting this issue as such recommends the higher-brain standard. I will take Robert Veatch's elaboration of this argument, which is the most developed and prominent in the literature, as representative.[55]

Veatch contends that the definition-of-death debate is primarily a moral issue:

[It] is actually a debate over the moral status of human beings. It is a debate over when humans should be treated as full members of the human community. When humans are living, full moral and legal human rights accrue. Saying people are alive is simply shorthand for saying that they are bearers of such rights.[56]

Surely that last sentence misses the mark. If it were true, then nihilists, who ascribe moral rights to no one, would deny that any of us is alive. Moreover, while the idea that human beings enjoy equal moral status is today widely accepted, perhaps especially in liberal Western democracies, many worldviews past and present reject the claim that all human beings have full, and therefore equal, moral status or rights. If Veatch were right, consider the implications of traditional Hindu beliefs about lower castes, Aristotle's view of "natural slaves," and some Americans' beliefs about homosexuals: Members of these groups are all dead! While these views may be morally obnoxious, they are hardly unintelligible and the people who hold them are not so confused about life and death.

More promising is another way Veatch articulates the thesis that the definition-of-death debate is a moral issue. What we are asking, he argues, is when we ought to discontinue certain activities, such as life-support efforts, and commence certain other activities, such as organ donation, cremation or burial, grieving, transfer of property, change of a survivor's

---

[54] This section draws significantly from my "Persons, Organisms, and Death," pp. 429–31.

[55] See also Gervais, *Redefining Death*, ch. 6. Gervais never convinces me that defining death is primarily a moral issue. Also, her suggestion that we view human death as "personal" death seems vulnerable to my criticisms of approaches that assume person essentialism or mind essentialism.

[56] "The Impending Collapse of the Whole-Brain Definition of Death," p. 21.

marital status, and so on.[57] To understand death, we must decide when *death behaviors* are appropriate and then classify those with respect to whom those behaviors are appropriate as dead.[58] And, of course, the question of when death behaviors are appropriate is a moral question.

Let's examine this claim. To say that someone has died does seem tantamount to saying that certain behaviors are now appropriate, while certain other behaviors are no longer. But this doesn't show that the statement that someone has died is *itself* a moral statement. It is sufficient to note that certain moral premises are assumed by virtually all participants in the debate. If, for example, everyone assumes that cremation or burial is appropriate only after one has died, a determination of death will seem tantamount to an assertion that it is now appropriate to cremate or bury. So the statement that someone has died might have moral importance, given other things we believe, without having moral content. Thus, Veatch's claim that the judgment that someone has died is itself moral is underdetermined by the sort of appeal just considered.

Further, his thesis is rendered somewhat doubtful by the reminder that the concept of death is at home in biology, which offers many instances in which a statement of death seems not to have moral implications. To say that a mosquito or clover has died has no clear moral significance. So it's implausible to hold that the general concept of death is moral. Might human death – or perhaps, more generally, the death of sentient beings – be a moral concept, even if the death of some organisms is not? Only if there is, in addition to the organismic concept, a moral conception of (or including) human death that deserves to take precedence. So far, we have found no reason to think so.

A more plausible assertion is that death is primarily a biological concept that, at least in the human case, is *morally very salient due to a relatively stable background of social institutions and attitudes.* We do not (with only ghastly exceptions) cremate or bury unless we believe someone is dead. We grieve when a loved one had died. To keep a dying person in the house often strikes us as kind; to keep a dead person there strikes us as repulsive. We discontinue medical treatment for someone who has died, though in recent decades we have decided that it is often permissible to forgo treatment for the living (note the qualification "*relatively* stable background"). Most discussions of death tacitly accept the dead-donor rule – that one must be dead for one's vital organs to be extracted for transplantation – so

---

[57] "The Whole-Brain-Oriented Concept of Death," pp. 15–16.
[58] Veatch (personal correspondence).

that proposed exceptions to this rule are generally considered radical. And so forth. Because of these institutions and attitudes, whether or not someone is dead matters to us morally.

It also merits stressing that discussions of the nature of death are frequently prompted by a moral or pragmatic agenda. So, if one thinks that the permanently unconscious are appropriate organ sources, one could avoid the perceived radicality of violating the dead-donor rule by convincing legislators to accept the higher-brain standard. Redefining death would obviate the judgment that harvesting organs from the permanently unconscious causes their deaths. And we have seen that much early interest in accepting the whole-brain standard was fueled by the practical agendas of facilitating transplants and avoiding futile treatment for certain patients. But none of this suggests that death itself is a moral concept.

Even if it were, it would hardly follow that the higher-brain standard is morally preferable to the traditional or whole-brain standards. Relatedly, we need to explain the *grounds* for moral judgments about death behaviors. A moral judgment of the form "Now it is appropriate to grieve [or the like]" invites the query "Why?" since moral judgments should be supportable by reasons. Now most people would answer, "Because he has died." But one cannot give that reply on the present account, since the answer – which simply means, supposedly, that such behaviors are now appropriate – would be redundant.

Perhaps Veatch and like-minded theorists would reply differently: "Because he has permanently lost the capacity for consciousness." But, to many people, it is not obvious why that answer explains the appropriateness of behaviors associated with death, or at least some of them, such as organ transplantation, cremation, and burial. One would need more of a justification to answer the "Why?" question. The best answer one could offer, I suggest, is that permanent loss of consciousness entails an existence that is devoid of value *for the unconscious individual herself.* That means that, even if death were a moral concept, a convincing case for the higher-brain standard would depend on an appeal to narrative identity or prudential value, as discussed in the previous section. And we found that such arguments cannot justify universal adoption of the higher-brain standard, but only an exemption from a default standard based on the organismic conception of death.[59]

---

[59] I believe that Veatch's moral case for the higher-brain approach ultimately rests on an ontological claim – that we are essentially human beings. Veatch formally defines death

WHERE SHOULD WE TAKE THE ORGANISMIC CONCEPTION?

We are essentially human animals. Like other organisms, we die on the permanent cessation of functioning of the organism as an integrated unit. But this organismic conception of death does not self-evidently support a specific standard of death for human beings. Which standard – cardiopulmonary, whole-brain, disjunctive, or some other – most faithfully specifies the organismic conception?

Against Brain Death

Despite the widespread acceptance of brain-death standards (whole-brain, brainstem, or disjunctive) in recent decades, I concur with the growing ranks of dissenters who believe that confidence in brain death has been misplaced.[60] According to the organismic conception, human death involves the permanent loss of functioning of the organism as an integrated unit – or "loss of integrative functioning" for short. The most important problem is that whole-brain death – and a fortiori brainstem death – are insufficient for loss of integrative functioning.

Neurologist Alan Shewmon has collated impressive evidence supporting this claim.[61] Many of our integrative functions, he argues, are not mediated by the brain and therefore can persist in patients who have been declared brain-dead by standard clinical tests. Such somatically integrating functions include these: homeostasis; elimination, detoxification, and recycling of cellular wastes; energy balance; wound healing; fighting of infections; and cardiovascular and hormonal stress responses

as the irreversible loss of whatever characteristics are essential to an entity (*Death, Dying, and the Biological Revolution* [New Haven: Yale University Press, 1976], p. 25) and asserts that "the essence of the human being is the integration of a mind and a body" ("The Impending Collapse of the Whole-Brain Definition of Death," pp. 21–2). This implies that permanent loss of mind, the capacity for consciousness, entails death for human beings – the thesis of mind essentialism. (Of course, the thesis that we are essentially minded is consistent with the thesis that we have other essential features, such as embodiment, as in Veatch's view.) I have already argued against this thesis.

60 See, e.g., Bartlett and Youngner, "Human Death and the Destruction of the Neocortex"; Veatch, "The Impending Collapse of the Whole-Brain Definition of Death"; Baruch Brody, "How Much of the Brain Must Be Dead?" in Younger et al., *The Definition of Death*: 71–82; Shewmon, "The Brain and Somatic Integration"; Potts, "A Requiem for Whole Brain Death"; Amir Halevy, "Beyond Brain Death?" *Journal of Medicine and Philosophy* 26 (2001): 493–501; Stuart Youngner and Robert Arnold, "Philosophical Debates About the Definition of Death: Who Cares?" *Journal of Medicine and Philosophy* 26 (2001): 527–37; and McMahan, *The Ethics of Killing*, pp. 426–42.

61 See, e.g., Shewmon, "The Brain and Somatic Integration," pp. 467–71.

to unanesthetized incisions (for organ retrieval). In several cases, brain-dead bodies have even proved capable of gestating a fetus, sexual maturation, and/or growth in size. Moreover, there are functions mediated by the spinal cord, such as certain reflexes. (To hold that the latter functions do not confer life, whereas any brainstem functions do, is, in effect, to draw a line – arguably, an arbitrary one – within the nervous system between the top of the spinal cord and the bottom of the brainstem.[62]) According to Shewmon, *most* somatically integrative functions are not mediated by the brain, so it would be better to describe the brain as an enhancer, rather than an integrator, of somatic functions. He marshals powerful evidence that most brain functions typically cited as integrative merely sustain an already presupposed unity.[63] Thus, neither respiration, circulation, nor nutrition is reducible to some integrative activity of the brain. Michael Potts elaborates on two of these three examples:

The most important of these integrative functions are respiration and circulation, since their continued functioning in an organism is necessary and sufficient for the life of the organism, from the level of cells to [that of systems]. Respiration, in the sense of the exchange of oxygen and carbon dioxide in the body, continues in brain dead patients; all the ventilator does is expand the diaphragmic muscles and provide oxygenated air. The circulatory system also continues to function in whole brain death, as evidenced by continued heartbeat, and in many patients, a stable blood pressure sustained without the need for pressor drugs. Such systemic functions often continue for a long period of time; long before Shewmon's recent work, it was known that rapid asystole [absence of cardiac contractions] was no longer an inevitable result of "whole brain death."[64]

Several scholars have noted a further source of embarrassment to proponents of the whole-brain standard: Many patients who meet standard clinical tests for whole-brain death continue to exhibit minor brain functions, such as isolated nests of perfused (living) brain cells, continuing electroencephalographic activity, and hypothalamic functioning.[65] The last, which controls neurohormonal regulation, is especially important, because it is clearly an *integrating* function of the brain.[66] Anyway, we are apparently not supposed to take the law literally when it refers to "all functions of the entire brain" in articulating the whole-brain standard.

---

[62] Veatch, "The Impending Collapse of the Whole-Brain Definition of Death," p. 21.
[63] Shewmon, "The Brain and Somatic Integration," p. 464.
[64] "A Requiem for Whole Brain Death," p. 484.
[65] See, e.g., ibid., p. 482; Veatch, "The Impending Collapse of the Whole-Brain Definition of Death," p. 18; and Youngner and Arnold, "Philosophical Debates about the Definition of Death," pp. 529–30.
[66] Brody, "How Much of the Brain Must Be Dead?" p. 73.

Interestingly, a clinical consultant for the President's Commission, James Bernat, has responded to such data about residual brain functions in brain-dead patients by revising his definition of death to "the permanent cessation of the *critical* functions of the organism as a whole."[67] This has permitted him to revise the standard of death to permanent cessation of the *critical* functions of the whole brain while denying that the residual brain functions are critical. According to Bernat, the critical functions of the organism as a whole are (1) the vital functions of spontaneous breathing and autonomic circulation control, (2) integrating functions that assure the organism's homeostasis, and (3) consciousness; a patient is dead on losing all three.[68] Yet, in their classic 1981 paper that helped to establish the conception of death underlying the whole-brain standard, Bernat and colleagues offered neuroendocrine control as their first example of the functioning of the organism as a whole![69] Why do Bernat and other whole-brain theorists find it so trivial today? I submit they are going to great, and sometimes implausible, lengths to preserve an appearance of legitimacy for the whole-brain standard.

Couldn't we modify practices of testing so that they included a test for neurohormonal regulation, which is clearly a type of integrative functioning? We could, but then we would have to abandon anything resembling current practices of organ transplantation (assuming we retain the dead-donor rule). Hormonal levels due to residual neurohormonal functioning sometimes remain constant for more than three days, dooming organs to unviability.[70]

So far, we have found two problems with the idea that human death is, or is entailed by, brain death: (1) brain death is not sufficient for loss of integrated functioning and (2) patients standardly classified as brain-dead by clinicians often do not satisfy the legal requirement (if we take it literally) of permanent cessation of *all* brain functions. A third problem is that brain death is not necessary for loss of integrated functioning. Suppose an entire brain is transplanted from one human organism, *A*, to another, *B*. The brain remains alive, now in *B*, and retains the capacity for consciousness. Suppose organism *A*, now brainless, is not provided life supports. *A* will quickly lose integrated functioning and qualify as dead

---

[67] "A Defense of the Whole-Brain Concept of Death," *Hastings Center Report* 28 (March–April 1998): 17.

[68] Ibid.

[69] "On the Criterion and Definition of Death," p. 390.

[70] Brody, "How Much of the Brain Must Be Dead?" p. 77. Brody cites H. J. Gramm et al., "Acute Endocrine Failure After Brain Death," *Transplantation* 54 (1992): 851–7.

by any criterion. So, in principle, loss of integrated functioning can occur even though brain death has not.[71]

Moreover, since brain-death theorists claim that the brain must integrate the organism's functioning for a human to be alive, their theory is challenged by the *locked-in syndrome*, as described by Bartlett and Youngner:

> The locked-in patient… suffers significant, but incomplete, destruction of the brain. [Unlike in PVS], the brain portions responsible for consciousness and cognition are intact. All portions of the brain stem and deep cerebral areas responsible for integration of vegetative functions have been destroyed. However, the blood supply and neural connections to the other cerebral areas, as well as the reticular activating system located in the brain stem, have been spared. Although the patient cannot spontaneously regulate respiration, blood pressure, temperature, hormonal balance, and other functions, he is awake and alert…. [His] life can only be maintained through the full efforts of the Intensive Care Unit staff.[72]

Everyone agrees that the conscious locked-in patient is alive. Because parts of the brain remain functional, the whole-brain standard delivers this verdict. So does Bernat's revised whole-brain standard, which understands death in terms of loss of *critical* functions, since he counts consciousness as a critical function. The problem that locked-in patients pose for brain death is that they exhibit no more integrated functioning than some brain-dead bodies do.

Although locked-in patients' brains function enough to support consciousness, they do not integrate somatic functioning. For these patients cannot move their body parts, except in some cases the eyes, and they cannot receive sensory information from most of their body. (In the most extreme cases, paralysis and lack of sensory processing are total.) Moreover, the brainstem's regulatory capacities, such as spontaneous respiration, are destroyed. With considerable life-support efforts, the locked-in patient can exhibit integrated functioning but, importantly, the brain isn't doing the integrating and the degree of integrated functioning may not exceed that of a brain-dead body.

Consider, for example, the brain-dead woman whose somatic functioning is sufficiently integrated to continue pregnancy. With external support, she is breathing, maintaining circulation and directing blood to the fetus, metabolizing, and so on.[73] According to the brain-death

[71] McMahan, *The Ethics of Killing*, p. 429.
[72] "Human Death and the Destruction of the Neocortex," pp. 205–6.
[73] In this paragraph I benefit from a discussion in McMahan, *The Ethics of Killing*, pp. 431–5.

theorist, she is dead while the locked-in patient is alive. Is this plausible? Admittedly, the locked-in patient's brainstem retains *some* functions, such as the reticular formation's making consciousness possible, but none that are *somatically* regulating. (I assume that somatic regulation involves more than regulation of another brain part.) In terms of somatic regulation, extreme cases of the locked-in syndrome and the case of the pregnant brain-dead woman are on a par.

Naturally, the brain-death theorist will stress the fact that the locked-in patient is conscious. But, since consciousness does nothing here for somatic regulation, this move will support the brain-death approach only if we abandon the requirement that the brain integrate the organism's functioning and accept a Bernat-like view that human life simply requires the presence of one critical function, such as consciousness. But that provokes the question of exactly what sort of conception of death could underlie this revised whole-brain standard.

Presumably, Bernat and like-minded thinkers intend to remain within the organismic approach. Holding that we are essentially (human) organisms, I concur with the general organismic approach. But I doubt the coherence of Bernat's revised specification of it, a formulation that counts consciousness as a critical function along with spontaneous respiration, circulation, and certain integrating functions – and on that basis identifies brain function as crucial. Consciousness is not crucial to organismic functioning, even for human beings. The pregnant brain-dead woman's body is functioning fairly well (with external support), and in an integrated way, though she isn't conscious. While her brain is not directing the integrated functioning of her body, that is also true of the undeniably living locked-in patient. That the latter is conscious is irrelevant to the issue of integrated functioning.

If consciousness really is a crucial function, then the view of our essence that motivates the conception of death is not really the organismic (biological) one. Rather, it is a hybrid of psychological and organismic conceptions: We are essentially beings who are minded, who maintain organismic functioning, or both.[74] But the coherence of this conception is uncertain. What if Vicky's whole brain is successfully transplanted while her original body is maintained with life supports? If she is essentially minded *or* organismic, then perhaps Vicky will persist qua *minded* in the body that receives her functioning brain and will persist qua *organismic*

74 Cf. ibid., p. 435.

in the original body. Then Vicky will be identical to two persons after the surgery, an impossible result.[75]

Another possible interpretation is that this scenario involves fission, in which case Vicky does not survive the surgery because neither resulting individual is she. This interpretation would require amending the present conception of our essence to include a requirement of uniqueness: that a human individual have only one *continuer* – one individual who is spatiotemporally continuous with her while satisfying the disjunctive criterion of being minded or maintaining organismic functioning – for the original human to survive. Perhaps that amendment would be plausible. But even if this conception is coherent and plausible as a view of our essence, it does not support the whole-brain criterion. Rather, it supports some variant of the traditional view, because, once again, brain-dead bodies often retain significant integrated functioning, no less than locked-in patients do. This result is certainly not what Bernat and like-minded theorists seek.

Thus it appears that the whole-brain standard, whether taken to include all functions of the entire brain or simply the brain's critical functions, is untenable. Nor does the UDDA's disjunctive standard or the brainstem standard fare better. The former wrongly implies that whole-brain death is sufficient for human death, so it inherits the basic problems of the whole-brain standard. The latter, meanwhile, wrongly implies that brainstem death is sufficient for human death. All brain death standards fail. Clinically, with the possible exception of Bernat's revised whole-brain standard, they overestimate the importance of the brain in integrating or regulating organismic functioning. Conceptually, it is uncertain whether any of them corresponds to a coherent view of our essence. Because the higher-brain view of death is also untenable, as discussed earlier, we are left with the perhaps surprising conclusion that the traditional understanding of the nature of death was closest to the mark.

Updating the Traditional View

According to the traditional standard, as formulated earlier, human death is the permanent cessation of cardiopulmonary function. Unlike the higher-brain and whole-brain standards, this formulation correctly implies that any human body that is breathing and maintaining circulation

---

[75] As discussed in Chapter 2, I assume both that a person can't have two bodies simultaneously and that identity is a transitive relation. Since person *A* and person *B* are distinct, Vicky can't be identical to both.

is alive, even if the continuation of these functions requires external support (as with brain-dead bodies, normal embryos, and locked-in patients). What is incorrect, or at least seriously misleading, about this formulation is its tendency to suggest that the difference between human life and death is determined by the functioning of two organs – heart and lungs – an overly reductionistic picture that neglects the holistic nature of bodily functioning.

Contrary to this reductionistic picture, integrative unity, as Shewmon stresses, exists diffusely throughout a living organism and need not have a single all-important regulator: "Integra*tion* does not necessarily require an integrator, as plants and embryos clearly demonstrate. What is of the essence of integrative unity is neither localized nor replaceable – namely the anti-entropic mutual interaction of all the cells and tissues of the body, mediated in mammals by circulating oxygenated blood."[76] From this perspective, the brain, like the heart and lungs, is a very important part of the mutual interaction among body systems, but it is not a sine qua non of life. Nor are the other organs and bodily systems passively dependent on the brain for their functioning. The brain can augment or refine these systems, but their preexisting capacity to function is necessary for the functioning of the organism as a whole. This is so even of a brain function as somatically integrating as the maintenance of body temperature – for which the "thermostat" is present in the brain, while the "furnace" is the energy metabolism diffused throughout the body. A house with a thermostat can optimize temperature control, but without a thermostat, a house that has a furnace can stay warm. Similarly, brain-dead bodies on life supports grow colder, if not covered with blankets, but not comparably to corpses.[77]

While a realistic picture of organismic functioning is holistic, certain functions are critical. The traditional view correctly identifies respiration and circulation as especially crucial. Now respiration is not simply the working of lungs, nor is circulation just the operation of the heart; both organs can be artificially replaced as the organism maintains integrated unity. Respiration and circulation are diffuse throughout the body. And they are crucial because they permit oxygenated blood to circulate to different bodily systems, a condition that is necessary and sufficient for the organism as a whole to function as an integrated, energy-preserving unit, resisting the entropy characteristic of inanimate matter. Unlike brain

---

[76] Shewmon, "The Brain and Somatic Integration," p. 473.
[77] The analogy is from ibid., p. 471.

death, loss of respiration and circulation leads inexorably to the dysfunction of cells, tissues, organs, and bodily systems – and therefore of the entire organism. Except in anomalous contexts such as profound hypothermia or cryopreservation (freezing of a body), the systemic "point of no return" occurs perhaps twenty or thirty minutes after cardiac arrest without resuscitation.[78] Thus, an updated traditional standard – which we can call *the circulatory-respiratory (CR) standard*[79] – asserts that human death is *the permanent cessation of circulatory-respiratory function*. What clinical tests are most appropriate for diagnosing death using this standard is a question we may leave for clinical experts.

Why Not Disaggregate Death into a Process?

Throughout this chapter, the discussion has assumed that we should attempt to determine which standard of death best captures the nature of death, as expressed in a plausible definition. I have argued that we should conceptualize death in organismic terms and that the CR standard is most consonant with such a definition. But must we choose only one standard of death? Perhaps different standards are appropriate for different purposes. If the boundary separating life from death is insufficiently clear and determinate to vindicate a single standard, the search for *the* correct standard of death may be futile precisely because it misconstrues the nature of death, taking it to be a determinate event rather than a process.

According to a proposal developed by Amir Halevy and Baruch Brody, we should disaggregate death into a process.[80] The nearly simultaneous emergence of ventilators and organ transplantation, the authors note, provoked three practical questions: (1) When may a doctor unilaterally discontinue treatment? (2) When may we take organs for transplantation? (3) When is a patient dead for legal purposes and in need of an undertaker? We have generally assumed that the same point in time must answer all three questions, but after more than thirty years of debate, "no one has succeeded in creating a unifying definition and criterion of death which

[78] See ibid., p. 469; and Potts, "A Requiem for Whole Brain Death," pp. 487–8.

[79] Shewmon uses the same term ("The Brain and Somatic Integration," p. 469). To its credit, the President's Commission also refers to "circulatory and respiratory functions," rather than "cardiopulmonary function" (which emphasizes the two organs), in proposing the disjunctive standard.

[80] Their seminal paper is Amir Halevy and Baruch Brody, "Brain Death: Reconciling Definitions, Criteria, and Tests," *Archives of Internal Medicine* 119 (1993): 519–25.

is [sic] internally consistent, consonant with known physiology, and does [sic] not eliminate organ transplantation from the clinical armamentarium."[81] (I challenge this claim with my own view.) A better approach, according to Halevy and Brody, is to answer each question on its merits, disaggregating death accordingly.

This, then, is their proposal.[82] First, doctors may unilaterally discontinue treatment, even over a family's objections and a patient's previously expressed wishes, when the patient has permanently lost the capacity for consciousness. The only exception is when the patient or family will bear the full costs of continued treatment. This policy represents a reasonable balance, the authors contend, between preserving limited resources and respecting the wishes of patients and their families. Second, organs may be removed for transplantation when the patient meets the whole-brain standard. Preserving the status quo, this policy strikes a balance between the desirability of viable organs and the preservation of public trust in organ transplantation (which would be significantly threatened by organ retrieval from PVS patients and anencephalics). Third, a patient may be considered dead, with the attendant social and legal implications, on meeting the traditional standard of death. If the patient's organs are removed for transplantation, she will meet this standard almost immediately. In other cases, if the patient is in PVS, cessation of nutrition and hydration will ensure death within two weeks, whereas if the patient is brain dead, removal of a respirator will ensure death within minutes.

The Halevy–Brody proposal represents one way to address the preceding three questions with plausible answers that will sometimes represent different points in time. Another way would be to answer the third question with the CR standard while abandoning the assumption that unilateral discontinuation of treatment and organ transplantation must await death (however understood). So why prefer their solution, which rests on the claim that death is a process? Brody clarifies that much of the rationale underlying their approach is

the fundamental insight of fuzzy logic, namely, that the world does not easily divide itself into sets and their complements. Death and its complementary property life determine mutually exclusive but not jointly exhaustive sets. Although no organism can fully belong to both sets, organisms can be in many conditions (the very conditions that have created the debates about death) during which they do not fully belong to either. That is why you cannot find the answers to the

---

[81] Amir Halevy, "Beyond Brain Death?" *Journal of Medicine and Philosophy* 26 (2001), p. 499.
[82] Here I draw mainly from Halevy, "Beyond Brain Death?" pp. 499–500.

questions by finding the right moment in the process to serve as the moment for belonging to the set of the dead. Death is a fuzzy set.[83]

Is death really a fuzzy set? If so, does this fact support the authors' proposal?

Unquestionably, the boundaries of death are somewhat blurry. Even if we confidently accept the CR standard, for example, we would lack a precise answer to the question of when the cessation of functioning should be considered irreversible or permanent. So any decision about precise clinical criteria – say, that twenty-five minutes with no circulatory or respiratory function is enough – will be somewhat arbitrary. But it hardly follows that no *standard* is correct. A significant assumption behind the Halevy–Brody proposal is that there is no uniquely correct answer to the question of our essence and therefore no uniquely correct definition of death.[84] I reject this assumption. On the basis of the biological view, defended in Chapter 2, I have argued that we are essentially human animals, so the organismic definition of death is correct. I have also argued that, in light of clinical facts and conceptual considerations, this definition is best specified with a CR standard. The reason there is so much controversy over standards of death, I suggest, is not that death is so fuzzy or vague a concept that it is more processlike than eventlike. The reason, rather, is epistemological: Participants in the debate often accept a standard of death for less than compelling reasons, without full awareness of the salient clinical and philosophical issues.

But suppose I am wrong. Suppose death is as vague a concept as adulthood. Just as the late teens and early twenties represent times when one is neither fully a child nor fully an adult, someone who is in PVS or is brain-dead but still breathing is neither fully alive nor fully dead. Even if this were true, there would be powerful practical incentives for drawing some reasonable line for declaring death. A great deal of confusion would likely result if we said things like "Grampa is partly dead, but less dead than Gramma, although she's not fully dead." Indeed, Halevy and Brody nearly suggest as much when they reserve the appellation "dead for social and legal purposes" for those who meet the traditional standard. People are used to thinking of life and death as mutually exclusive, exhaustive sets, so there would be practical advantages to drawing a sensible line that demarcated death this way.

---

[83] "How Much of the Brain Must Be Dead?" p. 72.

[84] Thus Brody pluralistically describes various definitions without suggesting that any is preferable to the others (ibid., pp. 72–3).

But why not draw several lines for death, as the authors in effect recommend: one at PVS, one at brain death, and one at asystole? After all, while practical considerations encourage line-drawing with respect to adulthood, we draw *several* for policy purposes: at ages sixteen, eighteen, and twenty-one at least. On the other hand, no one is confused by the idea that one can be an adult for some purposes but not others, or partly an adult, because one presumably survives the transition from full childhood to full adulthood. By contrast, people generally presume that one goes out of existence (at least in this world) on dying, a rather momentous change with far-reaching social and legal ramifications. Confusion here is both more disconcerting and more likely, since the idea of someone's only partly existing borders on the unintelligible. There seem to be several advantages and no significant disadvantage to drawing a single line for death, even if death is a process.

Thus, one might welcome the spirit of the authors' practical suggestions – drawing different lines in answering distinct practical questions – while rejecting their claim that death is a fuzzy set and the implication that PVS patients and brain-dead patients are partly, and to different degrees, dead. On the basis of arguments presented in this chapter, I do reject this claim and its implication. Human death should be conceptualized in organismic terms and specified by the CR standard.

SOME COMMENTS ON POLICY OPTIONS

The major aims of this chapter have been achieved. We have addressed the nature of human death, both at the conceptual level and at that of standards. We have also explored the role of personal identity theory in elucidating the nature of death. We found that the strongest theory of our numerical identity and essence, the biological view, supports the organismic conception. Meanwhile, concerns about the psychological view of our identity – and, more specifically, the variants associated with person essentialism and mind essentialism – undermined some of the leading arguments for the higher-brain standard. Moreover, appeals to narrative identity and what matters in survival failed to vindicate this standard but suggested the permissibility of exemptions from a default standard of death for those who autonomously choose the higher-brain standard in their own case. Investigation of the organismic conception in light of certain clinical facts favored the CR standard – an updated traditional standard – over the various brain death standards: whole-brain, brainstem, and disjunctive.

These findings suggest that the legal acceptance of brain death represents an abandonment of our shared biological concept of death. Contrary to what the President's Commission claimed, brain death is not simply one manifestation of what we have always regarded as death. Acceptance of brain death amounts to a *reconstruction* of this concept, a *new definition* of human death. This reconstruction, furthermore, is driven by pragmatic considerations and rationalized by a combination of clinical fudging and unpersuasive philosophical arguments. From a purely philosophical standpoint, then, the reconstruction is unjustified; the arguments for departing from the original concept of death are inadequate. Thus, strictly speaking, brain death is not a form of death, from which it follows that many people have been prematurely declared dead, some of them killed to procure their organs.

Then again, a very small percentage of the public is likely to engage in or care much about the subtle distinctions and sophisticated arguments that support our philosophical conclusions. Most of the public's concerns about the definition of death are likely to revolve around the issue of optimal public policy. Yet it is impossible to infer optimal policy directly from our philosophical findings. In concluding this chapter, I will simply outline two broad options for addressing the policy issues. Both options strike me as reasonable, yet neither is without difficulties and clearly superior to the other.

A Daring Policy Approach

The first approach is relatively daring. It insists on "calling a spade a spade," rejecting the classification of brain death as a form of human death since, strictly speaking, it is not. Accordingly, this option requires changing the law in all jurisdictions that have accepted brain death (whether whole-brain or brainstem death, whether exclusively or disjunctively with the cardiopulmonary standard). It is difficult to know how to achieve this end politically. Should government officials, citing several authors' or perhaps a new commission's work, apologetically announce that the legal acceptance of brain death had been a mistake, motivated by a President's Commission and other professional bodies who led the country astray? Although that approach is painful to contemplate, it is not obvious what a better strategy would be. However achieved, thorough legislative overhaul of standards of death would be necessary. Then again, to the extent that related laws – such as those concerning homicide, life insurance, and marital status – refer to the legal determinations of death without specifying what the standards for those determinations should be, these related laws would not automatically require revision.

Naturally, the present approach provokes a dilemma concerning organ transplantation. Either we change current practices such that organ procurement is permitted only after the CR standard has been met, dooming many or most organs to unviability, or we permit current organ procurement practices to continue despite the fact that they involve killing organ donors. My suggestion here would be to abandon the dead-donor rule, because (as I will explain in a moment) it is not morally required. We should continue to allow organ retrieval from patients who have met the whole-brain standard, assuming that, while competent, they provided a valid (e.g., well-informed, voluntary) consent to donate. Probably health professionals and policy experts will prefer to use such euphemisms as "Taking his heart will shorten his life" rather than saying "Taking his heart will kill him." But, in opposing mainstream practices of organ procurement, some individuals and groups will surely publicize the fact of killing. Does that likely development doom this policy option to failure?

To assume so would seem premature. Admittedly, given present sensibilities, it sounds very bad to say that transplant surgeons will kill organ donors. But the reason this sounds so bad, I think, is not that the surgeons' action would be wrong. Talk of killing organ donors sounds unpalatable because people (1) often conflate the meanings of *kill* (to cause death) and *murder* (to kill wrongly) and (2) tend to forget that the rule "Do not kill human beings" has several nearly universally accepted exceptions. These exceptions include killing as a last resort in self-defense and killing suffering, terminally ill patients with pain medication when death is a consequence that was unintended but foreseen as possible.[85] Other commonly (though somewhat less widely) embraced exceptions to the rule against killing human beings, even where killing is intentional and the individual killed is innocent, include cases involving suicide, physician-assisted suicide, voluntary active euthanasia, and abortion. We could preserve the policy status quo of procuring organs from brain-dead patients by arguing that this instance of killing in accordance

---

[85] Those who accept the doctrine of double effect might even argue that taking organs from brain-dead but living patients is permissible because the foreseeable death is not intended and is not a means to the intended good effect of procuring an organ that can save another's life. For an introduction to this doctrine, see, e.g., Peter Knauer, "The Hermeneutic Function of the Principle of Double Effect," in Charles Curran and Richard McCormick (eds.), *Readings in Moral Theology No. 1: Moral Norms and Catholic Tradition* (New York: Paulist Press, 1979): 1–39.

with patients' prior autonomous wishes is a justified form of active euthanasia. Apparently, many citizens of Japan and Germany doubt that
brain death is death, yet believe organ transplantation is appropriate on
a declaration of brain death,[86] suggesting their willingness to make an exception to the dead-donor rule. Perhaps the citizens of the United States
and the United Kingdom, for example, would accept the demise of this
rule more readily than one might suppose.

The more daring policy option, then, would change laws about when
death can be declared while preserving the status quo concerning organ
transplantation by classifying the killings involved as justified instances
of active euthanasia. Importantly, this approach does not entail keeping
brain-dead patients on life supports just because they are alive. Through
the use of advance directives, patients may prospectively authorize the
removal of life supports, including nutrition and hydration, on a determination of brain death – or on entering PVS, permanent coma, or some
other specified medical condition. (Chapter 5 will discuss advance directives in detail.) As for patients who left no such directives, treatment
may often be discontinued in accordance with the substituted judgment
standard or the best interests standard. Brain-dead patients will die a
few minutes after life supports have been removed. Patients who are
permanently unconscious but not brain-dead will die within a week or
two of the discontinuation of nutrition, hydration, and any other life
supports.

In some cases in which withdrawal of life supports is not authorized
by an advance directive, substituted judgment, or assessment of the patient's best interests, health professionals and their institutions may refuse
to provide further care on grounds of medical futility, understood as
the impossibility of providing medical benefit to the patient. Halevy and
Brody favor permitting unilateral withdrawal of care on a determination
of permanent unconsciousness, except when the patient or her family is
able and willing to pay for care. While this policy may prove problematic
in certain situations – for example, when family members need time to
arrive at a hospital and spend time with the unconscious patient, yet cannot pay for the extended care – health care institutions and individual
practitioners would do well to clarify their standards for what they regard
as justified withdrawal of care. Health institutions should also promote
the use of advance directives, the use of which will often obviate unwanted
end-of-life treatment.

---

[86] See the citations in footnote 22.

Conscience-Based Exemptions – on Either Approach

My recommendation regarding conscience-based exemptions to the default standard of death applies to both of the policy approaches I am describing. The CR standard proved to be the best specification of the organismic conception of death, and no argument in favor of reconstructing human death across the board proved persuasive. The appeal to narrative identity and prudential value, however, provided grounds for allowing autonomous individuals to reconstruct the meaning of death in their own cases. This constitutes one basis for permitting individual exemptions.

Another basis rests on considerations of our numerical identity and essence. Assuming the arguments of Chapter 2 are correct, the biological view is stronger than any psychological view. But that hardly entails that all variants of the psychological view are completely unreasonable. Mind essentialism, I think, deserves our respect, after all the arguments are in, even if the most judicious evaluation of the arguments vindicates the biological view. At issue here are the rights of a philosophical minority whose view of human identity is within reasonable bounds. Public policies should not rest solely on the most defensible understanding of the truth, especially where more or less reasonable disagreement persists. Because mind essentialism supports the higher-brain approach, we should permit individuals to exempt themselves from the default standard of death and embrace the higher-brain standard for themselves.

Is there any basis for permitting exemptions from the default standard of death in favor of the whole-brain, brainstem, or disjunctive standard? I am not aware of any religious convictions that single out one of these standards as the correct account. And given the conceptual problems associated with these standards, I am not aware of a coherent philosophical position that would do so. On the other hand, it would be very awkward politically and socially to permit exemptions in favor of the higher-brain standard but not in favor of any of these alternatives. After all, these alternative standards have been enshrined in the laws of many countries. Besides, even if there is no coherent philosophical or religious view that favors the idea of brain death as death, an individual may have personal reasons for favoring brain death in her own case. Perhaps she is a cognitive neuroscientist who identifies strongly with her brain in both its cerebral and brainstem capacities.

Thus I recommend allowing competent adults to exempt themselves from the CR standard in favor of any of the other widely recognized standards: higher-brain, whole-brain, brainstem, or disjunctive. I doubt that much greater liberality would be wise, however. So, for example, I do not recommend allowing individuals to adopt a personhood standard stating that they shall be treated as dead as soon as they lose the capacity for complex forms of consciousness (say, on becoming severely demented). To treat a still-conscious human being as dead simply seems too great a departure from our original concept of death and too inviting of abuse of vulnerable individuals who are close to meeting this criterion; its vagueness is a further major disadvantage. Also inadvisable, I think, would be allowing someone to adopt a putrefaction standard in his own case. Putrefaction is clearly not a necessary condition for our demise, and permitting this standard would be highly offensive to the public's sensibilities. So probably only the widely recognized standards should be options for individual adoption.

This policy regarding conscientious exemptions has additional practical advantages. First, it is likely to alleviate the problem of extensive unwanted care toward the end of life. Also, prospective organ donors who are uncomfortable with abandoning the dead-donor rule could, by adopting one of the alternative standards, avoid this difficulty, for *their* deaths will occur before organs are removed.

Of course, any detailed development of this approach will need to address the question of whether parents or guardians may make exemptions on behalf of their dependents. It will also need to specify appropriate requirements for ensuring valid consent to an exemption. I will not take up these issues here.

A More Cautious Policy Approach – and Conclusion

Another approach, which I find at least as promising as the first, would stress the confusion and awkwardness that would result from changing laws about death standards. According to this approach, the untruth underlying current policy – that brain death is a coherent specification of the traditional biological understanding – is largely tolerable. While some philosophical purists and some moral conservatives will continue to chafe at this pragmatically driven legal status quo, most people will not. In addition to avoiding the need to change laws about death, the status quo has the advantage of continuing relatively efficient practices of organ transplantation without creating the widespread social impression that they violate the dead-donor rule (although purists will disagree). Moreover, the status quo provides an opening to expand the range of personal exemptions from the default disjunctive standard. Currently, New York and New Jersey permit individuals to adopt, for religious reasons, the traditional standard as the only criterion for death in their case. States, or perhaps the federal government, could mandate that all of the major standards become options for conscientious adoption, and not only on religious grounds. Meanwhile, professionals and institutions could do more to clarify the criteria by which they will no longer be willing to provide care near the end of life while promoting greater use of advance directives, thereby reducing conflict and, frequently, expense.

Those who prefer the second, less daring policy approach may find the payoff from personal identity theory to be relatively modest. At the policy level, the only recommended change would be to increase the range of permitted individual exemptions from the default standard of death. Those who favor the more daring approach may find personal identity theory to have yielded more practical fruit in the form of revisionary policy suggestions. Yet, both groups should appreciate that undercutting the rationale for unjustified policy changes – such as legislating the higher-brain standard for everyone – is an important, if unexciting, practical result. Equally importantly, we should not evaluate the payoff of philosophical inquiry entirely in practical terms, for much of our endeavor has had a distinctively philosophical aim: to determine the nature of human death while clarifying its relationship to human identity.

5

Advance Directives, Dementia, and the Someone Else Problem

Al recently retired as a high school physics teacher at the age of sixty-five. Throughout his working years, he remained single, never had children, and mostly devoted himself to teaching and to such hobbies as learning foreign languages, playing chess, and keeping abreast of developments in the natural sciences and artificial intelligence. Now he and his physician, Dr. Juana Pedemonte, suspect that he is in the early stages of Alzheimer's disease, a type of progressive dementia. Their suspicion is based on clinical symptoms and the frequent occurrence of this disease in Al's family. While still competent, and in the context of detailed discussions with his physician, Al completes an advance directive. He stipulates that, once he has become so demented as to be "unable to remember most of my life, unable to communicate effectively, and unable to perform most everyday life tasks," he should not receive life supports (other than nutrition and hydration) if he enters a potentially fatal medical condition. This preference holds, he elaborates, even in the event that Al appears contented while profoundly demented. For he considers any existence in such a mentally compromised state incompatible with how he sees himself; a protracted, deep dementia would seriously diminish his life as a whole, as an incongruous ending can ruin an otherwise good story. He and Dr. Pedemonte agree that this directive aptly expresses Al's self-understanding and values.

Ten years later, Al is severely demented: barely able to recognize family members, unable to speak, incapable of planning one minute into the future, and largely bereft of memory of his own past. Although Al is clearly sentient and retains a *dim* awareness of himself as someone enduring over time, his self-awareness, recognition of others, and ability to complete sequences of actions are probably inferior to – and certainly no greater than – those of an ordinary dog. In this state, Al lacks the complex forms of consciousness that constitute personhood (as discussed in Chapter 1). He is, in short, a sentient nonperson. When Al contracts pneumonia, his physician understands that Al will die if he does not receive antibiotics and that his advance directive clearly calls for withholding any such lifesaving intervention. But Dr. Pedemonte feels conflicted. While the directive is clear, and although both anticipated this possible scenario, Al now seems contented most of the time despite (or perhaps partly because of) his obliviousness. Is it right to allow a contented old man to die when a simple intervention would save him? For Al now has no awareness either of the directive or of life as the person who completed that directive. What authority does it carry in making decisions for Al in his present state? Dr. Pedemonte wishes she had examined this dilemma more thoroughly ten years ago.

Beginning with the famous 1976 Karen Quinlan case and continuing with a substantial list of frequently cited decisions, American courts have found that medical life supports may in some circumstances be withheld or, if treatment has begun, withdrawn from incompetent patients.[1] (We use the term *forgo* to cover both withholding and withdrawing of treatment.) The foundational legal premise is that competent patients have a right to refuse life-sustaining medical treatment, while countervailing state interests in preventing suicide, preserving human life, and maintaining the medical profession's integrity fail to justify overriding the patient's prerogative. Of course, incompetent patients are presently incapable of self-determination; by definition, they lack the capacity for autonomous decision making. The courts have assumed, however, that the right to refuse treatment can survive loss of competence;[2] in effect, they have endorsed the idea of *precedent autonomy* (a concept explored in depth in a later section). What is needed to make this right effective on behalf of an incompetent patient is a firm basis for asserting that, while competent,

---

[1] *In re Quinlan*, 70 N.J. 10, 355 A.2d 647 (1976). See also, e.g., *Superintendent of Belchertown v. Saikewicz*, 373 Mass. 728, 370 N.E.2d 417 (1977); *Eichner v. Dillon*, 52 N.Y.2d 363, 420 N.E.2nd 64 (1984); *In re Conroy*, 98 N.J. 321, 486 A.2d 1209 (1985); and *In re Jobes*, 108 N.J. 394, 529 A.2d 434 (1987).

[2] Rebecca Dresser and John Robertson lucidly present the courts' reasoning in "Quality of Life and Non-Treatment Decisions for Incompetent Patients: A Critique of the Orthodox Approach," *Law, Medicine, and Healthcare* 17 (1989), pp. 234–5.

her preference was not to have life-sustaining treatment should she enter a prolonged state of incompetence.

Representing one such basis are advance directives, of which there are two types. *Instructional directives* give instructions for what sorts of medical interventions to provide or withhold from the patient. *Proxy directives* designate a surrogate decision maker to stand in for the patient if she is incompetent. Since surrogates need a basis for making the decision that the patient would have preferred – assuming the patient had treatment preferences – surrogates can benefit from a clear instructional directive. This chapter will focus on instructional directives.

In cases where an incompetent patient had not completed an advance directive while competent, courts have generally appealed to the doctrine of *substituted judgment*, permitting a family member or another appropriate surrogate to attempt to determine what the patient would have wanted in the present circumstances.[3] If there is no reasonably clear basis for determining what the patient would have wanted, the law turns to the doctrine of *best interests:* Treatment decisions are to be determined by what appears to be in the patient's best interests.[4] Decision makers typically make substituted judgments or apply the best interests standard without any judicial proceeding. Ordinarily, courts are involved only where there are significant uncertainties about what the law requires in the situation at hand.

Two legal developments following the cases previously cited stimulated greater interest in advance directives. In 1990 the Patient

---

[3] It is difficult to articulate the guiding idea here, because we are asking what a patient would have wanted – on the basis of her autonomously held preferences and values – regarding a situation in which she is incompetent and nonautonomous. We may conceptualize what she would have wanted as (1) what she did in fact want, while competent, to be done to her should she become incompetent or (2) what she would have wanted, while competent, to be done to her while incompetent *if she had considered the matter.* Possibility (1) seeks an actual preference, one the patient had. Possibility (2) seeks a hypothetical preference, representing the most faithful extrapolation from her values and (actual) preferences to the circumstances under consideration. Although I will not pursue this conjecture here, perhaps a hypothetical preference is a type of actual preference – namely, a *dispositional* preference, one an individual is disposed to become aware of on considering the matter, even if in fact she has yet to do so. If this is correct, a contrast with preferences of type (1) is still possible, because they were, at some point, *occurrent* actual preferences, ones consciously formed.

[4] Here we assume that some individual or institution is willing to pay for any treatment that is in the patient's best interests. Unless otherwise stated, this assumption will hold throughout our discussion.

Self-Determination Act (PSDA) became law. It requires hospitals and nursing homes to offer information about advance directives to incoming patients. The other development, also in 1990, involved the U.S. Supreme Court's first decision regarding a right-to-die case (which affirmed lower courts' findings that competent patients have a right to refuse even life-sustaining medical treatment, and that artificial nutrition and hydration are forms of medical treatment).[5] Nancy Cruzan had entered a PVS as a result of a 1983 automobile accident. Cruzan's parents wanted to remove a gastrostomy (feeding) tube from Nancy, citing their understanding of her values. But Missouri law, set out by the Missouri Supreme Court in their ruling in this case, required the continuation of nutrition and hydration despite the apparent impossibility of Nancy's ever recovering consciousness. Missouri law sets a demanding standard to legitimate forgoing life-sustaining treatments under the doctrine of substituted judgment: "clear and convincing evidence" that the now-incompetent patient would have wanted treatment to be forgone. The U.S. Supreme Court judged that states adopting such stringent standards do not thereby violate patients' or their families' constitutional rights. The Court's ruling meant that, in some states, nothing less explicit than an advance directive would be effective in forgoing life-sustaining treatment.[6]

Consistent with these legal developments, most leading literature in biomedical ethics regards advance directives as carrying significant authority in decision making for incompetent patients. Advance directives are generally conceptualized in one of two ways. In what may be the most common understanding, they are considered an especially credible basis

---

[5] *Cruzan v. Director, Missouri Department of Health*, 110 S.Ct. 2841 (1990).

[6] In an earlier, less publicized case that did not go to the U.S. Supreme Court, a New York court also validated the clear and convincing evidence standard for a surrogate to reject life-sustaining treatment. If there is any significant doubt about what the patient would have wanted, the court ruled, decisions must be made to preserve life. Interestingly, Missouri proved to be one state where an advance directive would not always be necessary to meet the clear and convincing standard. After Nancy Cruzan's case became widely known, some old friends of hers realized that the Supreme Court case involved Nancy (whom they knew by a different last name). When these friends testified that Nancy had wanted to be allowed to die if she entered PVS, a lower court decided that the cumulative evidence favoring removal of the feeding tube was clear and convincing. So, five months after the Supreme Court decision, Cruzan's feeding tube was removed and she died. See Gregory Pence, *Classic Cases in Medical Ethics*, 2nd ed. (New York: McGraw-Hill, 1995), pp. 17–20.

for making substituted judgments.[7] The assumed hierarchy for decision making is as follows:

1. Informed consent (for competent patients).
2. Substituted judgment (for formerly competent patients where there is sufficient evidence regarding their treatment preferences).
3. Best interests (for all other patients, including minors).

Alternatively, advance directives may be understood as a basis for decision-making that is distinct from substituted judgments. On this view, to complete an advance directive is to perform an act of will and not merely to provide evidence of one's preferences. A deliberate choice – as expressed in a wedding vow, a will, an explicit act of consenting, or an advance directive – has greater moral weight than one's preferences or desires in the absence of any deliberate choice.[8] Hence the following hierarchy:

1. Informed consent.
2. Advance directive.
3. Substituted judgment.
4. Best interests.

Assuming the advance directive standard applies only when an advance directive is reasonably *clear* in its direction for the case at hand, and *valid* insofar as its author competently made an informed, voluntary choice, then, according to this model, it makes sense to prioritize this standard over those of substituted judgment and best interests. (In effect, similar conditions apply to the other standards to warrant their place in either hierarchy: One must competently make an informed, voluntary decision for it to count as an informed consent and therefore as authoritative; a substituted judgment must be sufficiently supported by evidence to enjoy priority over best interests considerations; and a best interests judgment must be reasonable to count as having any authority at all.)

Not surprisingly, the authority of advance directives is sometimes challenged in particular cases. Perhaps the patient wasn't really competent when she completed the directive. Or perhaps she was competent but

---

[7] See, e.g., President's Commission for the Study of Ethical Problems in Medicine and Biomedical and Behavioral Research, *Deciding to Forego Life-Sustaining Treatment* (Washington, DC: Government Printing Office, 1983), ch. 4.

[8] See, e.g., Allen Buchanan and Dan Brock, *Deciding for Others: The Ethics of Surrogate Decision Making* (Cambridge: Cambridge University Press, 1989), pp. 115–17.

did not envision the sort of circumstance she is now in and, had she done so, would have decided differently about the present case than a literal reading of her directive suggests. There is a well-developed literature that addresses these and other questions.[9]

Rather than focusing on these familiar questions, this chapter will address the authority of advance directives in cases of severe dementia primarily through the lens of personal identity theory, with a secondary emphasis on the concept of precedent autonomy. The focus on *severe* dementia is motivated by the way this condition raises special issues concerning our identity.

THE NONIDENTITY THESIS AND THE SOMEONE ELSE PROBLEM

The Reasoning

In the case that began this chapter, Al represents patients who are so (irreversibly) demented that they no longer possess the capacity for the complex forms of consciousness that constitute personhood.[10] For convenience, let us reserve the term *severely demented* for such sentient human nonpersons. This section will concentrate on a pattern of reasoning that can be reconstructed as follows:

(1) Al-at-sixty-five is a person.

(2) Al-at-seventy-five is not a person.

Ergo

(3) Al-at-sixty-five and Al-at-seventy-five are not *the same* person.

Ergo

(4) Al-at-seventy-five is numerically distinct from – specifically, a successor to – Al-at-sixty-five.

---

9 See, e.g., Norman Cantor, *Advance Directives and the Pursuit of Death with Dignity* (Bloomington: Indiana University Press, 1993); Jim Stone, "Advance Directives, Autonomy, and Unintended Death," *Bioethics* 8 (1994): 223–46; Joan Teno et al., "Do Formal Advance Directives Affect Resuscitation Decisions and the Use of Resources for Seriously Ill Patients?" *Journal of Clinical Ethics* 5 (1994): 23–30; Thomas May, "Reassessing the Reliability of Advance Directives," *Cambridge Quarterly of Healthcare Ethics* 6 (1997): 325–38; Thomas Mappes, "Some Reflections on Advance Directives," *APA Newsletters: Newsletter on Philosophy and Medicine* 98 (Fall 1998): 106–11; and Robert Olick, *Taking Advance Directives Seriously* (Washington, DC: Georgetown University Press, 2001).

10 For a detailed description of Alzheimer's disease, the most common form of dementia, see American Psychiatric Association, *Diagnostic and Statistical Manual of Mental Disorders*, 4th ed. (rev. ed.) (Washington, DC: APA, 2000), pp. 154–7.

(References to "Al-at-sixty-five" and "Al-at-seventy-five" keep open the question of whether the referents are one and the same individual or, as the argument claims, numerically distinct individuals.) This reasoning suggests that the person who wrote the advance directive is not the individual to whom it would apply. This creates a problem – which we call *the Someone Else Problem* – for the authority of advance directives in cases of severe dementia. That is because the advance directive one completes is supposed to authorize certain decisions regarding *one's own* future medical treatment, not someone else's.[11]

While much of the literature in this area is not very explicit, several authors apparently reason along the lines of the previous syllogism, concluding that the author of the directive and the demented nonperson to whom it legally applies are distinct individuals – the *nonidentity thesis*. For example, in an influential article, Allen Buchanan refers to the author of the advance directive as "formerly existing" and asks whether we should "ignore the advance directive (now that its author has ceased to exist)."[12] Although he ultimately judges the nonidentity thesis to be not much of a problem, this conclusion is based largely on claims regarding so-called *surviving interests* (to be discussed later), not on doubts about the nonidentity thesis: "A person who issues an advance directive may do so not only to exercise control over what happens to her*self* after she become incompetent, but also to protect certain interests she has in what happens to her body after she, the particular person, no longer exists."[13] Later, while stressing that the two individuals are intimately related, he continues to assert nonidentity: "This would amount to recognizing as morally significant the especially intimate relationship between these individuals while at the same time acknowledging that they are distinct individuals."[14] Later collaboration between Buchanan and Dan Brock features the same reasoning and very similar phrasings.[15] Buchanan and Brock also say, quite starkly, "Jones [the author of the advance directive] no longer exists" and speak frequently of the earlier person's "successor."[16] Several other authors reason similarly. Rebecca Dresser, for example, states that

---

[11] For an earlier discussion, see my "Advance Directives, Dementia, and 'the Someone Else Problem'," *Bioethics* 13 (1999): 373–91.

[12] "Advance Directives and the Personal Identity Problem," *Philosophy and Public Affairs* 17 (1988), pp. 285–6.

[13] Ibid., p. 286.

[14] Ibid., p. 298.

[15] *Deciding for Others*, pp. 162–6, 183.

[16] Ibid., pp. 164–6 (quote from p. 165).

"[u]nderlying the advance directive concept is the assumption that the person executing the instruction or proxy directive is *the same person* whose medical care will be governed by the directive" before challenging that assumption.[17]

Sometimes authors suggest openness to the reasoning leading to the nonidentity thesis rather than explicitly endorsing it. Thus, for example, Rebecca Dresser and John Robertson say this about Parfit's approach to personal identity, which they consider one of two leading options without deciding between them: "According to this theory, advance directives for non-treatment could be issued by a different person than the subsequently incompetent individual. In such a situation, the former self's preferences would have no particular authority to govern the incompetent patient's treatment."[18] Meanwhile, Jeffrey Blustein concedes that "if the purpose of a theory of personal identity is to provide an answer to the question [of numerical identity], then reflection on personal identity poses a potentially serious threat to the moral authority of advance directives."[19] That is because "the nonperson is someone else, a successor to the agent of the advance directive (if not a person), so a question still remains about how the advance directive should figure in treatment decisions for the profoundly demented individual."[20] As we will see later, Blustein contends that it is more fruitful to set aside the issue of numerical identity and take up that of narrative identity. Thus, some of his statements suggest acceptance of the nonidentity thesis, but not the inference that the authority of advance directives in cases like Al's is problematic. Providing a final example, Helga Kuhse concedes that the nonidentity thesis may be correct while denying that it poses a problem for advance directives: "those who argue that advance directives rest on a confused understanding of personal identity may well be correct, but [that] does not by itself provide strong reasons for overriding refusals of life-sustaining treatment."[21]

---

[17] "Advance Directives, Self-Determination, and Personal Identity," in Chris Hacker, Ray Moseley, and Dorothy Vawter (eds.), *Advance Directives in Medicine* (New York: Praeger, 1989), p. 157 (emphasis added).

[18] "Quality of Life and Non-Treatment Decisions for Incompetent Patients," p. 236.

[19] "Choosing for Others as Continuing a Life Story: The Problem of Personal Identity Revisited," *Journal of Law, Medicine and Ethics* 27 (1999), p. 20.

[20] Ibid., p. 30.

[21] "Some Reflections on the Problem of Advance Directives, Personhood, and Personal Identity," *Kennedy Institute of Ethics Journal* 9 (1999), p. 361.

As these samples from the literature demonstrate, quite a few authors in bioethics assert, or at least express openness about, the nonidentity thesis. Some also accept the alleged implication that this creates a problem for the authority of advance directives in cases of severe dementia. All agree that, if the author of the directive is not the same person as the severely demented patient, then the two are numerically distinct individuals. What are we to make of this reasoning?

Critique of This Reasoning

As it stands, the argument is clearly fallacious. It does *not* follow from "Al-at-sixty-five and Al-at-seventy-five are not the same person" that "Al-at-seventy-five is numerically distinct from Al-at-sixty-five." The conclusion of numerical distinctness requires an additional premise, namely, that Al-at-sixty-five is essentially a person and therefore cannot exist at any time without being a person at that time. Presumably, this premise rests on the more general thesis of person essentialism: We (who are now) human persons are essentially persons. In Chapter 2 we found person essentialism to be an unattractive thesis about our essence. Undermining person essentialism undermines the preceding reasoning. It is not the case that an advance directive's author and a later demented individual need to be the same *person* in order to be numerically identical, for we are not essentially persons. What I find most remarkable about this reasoning is that those advancing it routinely fail to notice that they are making a contestable assumption about our essence.

In Chapter 2, we noted some reasons to prefer *mind* essentialism – the thesis that we are essentially beings with the capacity for consciousness – over person essentialism, even if the thesis that we are essentially human animals proved the most defensible. What if our arguments have been partly mistaken and mind essentialism is the most defensible account? Importantly, this would not vindicate the previous reasoning. If we are essentially beings with the capacity for consciousness, then Al survives into severe dementia since the demented individual retains that capacity so long as he is sentient. (Mind essentialism entails that permanent coma and PVS are forms of death, but the reasoning under consideration concerns demented individuals – who are sentient, not permanently unconscious.)

Having developed an account of our essence and numerical identity earlier in this book, we have been able, quickly and straightforwardly, to expose the fallacy in the reasoning leading to the nonidentity thesis. The critique may be strengthened by noting other difficulties generated by this reasoning and the Someone Else Problem it supposedly motivates.

The Odd Implications of the Nonidentity Thesis

Consider the following. The nonidentity thesis asserts that the author of an advance directive, say, Al-at-sixty-five, is numerically distinct from the later severely demented individual, say, Al-at-seventy-five. This implies that Al-at-sixty-five literally went out of existence at the onset of severe dementia – that is, dementia so advanced that it destroys the capacities that constitute personhood. But most of us assume, almost irresistibly, that when human beings go out of existence (at least in this world), they *die*. If this common assumption is correct, then the nonidentity thesis implies that, despite bodily, biological, and minimal psychological continuity (e.g., continuing capacity for consciousness, retention of some modest cognitive abilities) between Al-at-sixty-five and Al-at-seventy-five, the former dies somewhere in that continuum – unknown to anyone, including the most attentive family members and nursing personnel.[22] This is an extraordinary implication.

No less extraordinary is the implication that a *new* human individual, Al-at-the-onset-of-severe-dementia, comes into being, again unknown to anyone. But how could a human individual come into being without ever having been a fetus, baby, or even a child? Was he "born" old? Or, supposing, for example, that severe dementia sets in at age seventy, should we believe that Al-at-seventy-five is really only five years old? Either possibility is bizarre.

The second possibility is not so strange, one might argue, in view of an analogy with dissociative identity disorder. When someone develops this condition, such that a distinct personality emerges and temporarily controls the "shared" body – sometimes causing observers to judge that they are dealing with a new person – perhaps it would be natural to regard the alternate personality's age as the number of years since its emergence. Now, as noted in Chapter 2, the interpretation of dissociative identity disorder is very controversial. For present purposes, it is sufficient to note that the case for saying that Severely Demented Al's age is determined by the amount of time since the onset of severe dementia is far weaker than the case for saying that Alternate Personality is at a certain age

[22] Olick also notes this implication of two deaths (*Taking Advance Directives Seriously*, p. 143).

determined by the time of its emergence. Whereas Alternate Personality may impress other people as a distinct individual with whom to interact socially, Demented Al, for two reasons, probably won't. First, the slide from cognitive normalcy to severe dementia is gradual and continuous, frustrating any attempt to assign a "birthday" that doesn't seem highly arbitrary. Second, severe dementia massively damages one's capacity for social interaction, the capacity that chiefly motivates others' assignment of a new personality in dissociative identity disorder cases. I conclude, then, that the oddity of saying that Al-at-seventy-five is, say, five years old is little diminished by familiarity with dissociative identity disorder cases.

Now, one could avoid the implication that a new human individual came into being when the other one "died" by asserting that throughout the person's existence there was another human individual – more precisely, a human animal – who preceded the person, overlapped or perhaps constituted him, and now continues to live in a severely demented state after the person's disappearance. But that implies that there are now two distinct substances with different persistence conditions writing this book: the human animal and the person.

"Is this so implausible?" one might ask. "If I am a father and a Unitarian, can't I correctly say that a human animal is writing this book, a father is as well, and so is a Unitarian?" But the analogy doesn't hold, because there is no suggestion that *father* or *Unitarian* is a substance concept. Recall from Chapter 2 that a substance concept provides the most basic answer to the question "What kind of thing is X?" And no one holds that one who is a father has never existed, and cannot exist, at any time without being a father at that time. By contrast, surely *human animal* (or some broader biological category such as *animal*) is a substance concept; surely a presentient fetus is most fundamentally, or essentially, a human animal and will exist just so long as the animal does. The controversial claim, which I reject, is that *person* is also a substance concept. If it is, there are two substances writing this book – despite my having no interest in coauthorship. (And which one am *I*?)

Note that the two-substances hypothesis is not merely odd on its face. It also undercuts the nonidentity thesis by implying that Al-at-sixty-five and Al-at-seventy-five are the same human animal even if not the same person, in which case the severely demented patient is *not* someone else. One might reply that it was the *person* Al who wrote the directive, the *person* Al who is one of us (and, like us, essentially a person), while the demented Al, the rather old human animal, is indeed someone else. But that won't do. For on the two-substances hypothesis, the human animal, who eventually becomes that demented patient, was there all along. Since he and the person share the hand that writes, the human animal also signed the advance directive, just as the human animal, who shares a functioning brain with the person, was competent to give voluntary, informed consent. It appears, then, that the cost of avoiding the bizarre implications that a new human individual came into being at the onset of dementia is pulling the rug from under the nonidentity thesis – as well as generating odd implications such as the two-substances thesis and the problem of explaining cogently the person–human animal relationship (see Chapter 2).[23]

So it looks as though the nonidentity thesis implies that the advance directive's author dies at the onset of severe dementia as a new individual comes into being. Regarding the alleged implication of an invisible death, however, one might reply as follows: "We can appropriately avoid the term *death*, whose biological associations are out of place here, while asserting that the advance directive's author goes out of existence. True, we normally assume that, when one of us goes out of existence (at least in this world), she dies, but perhaps a compelling theory of our identity and essence could persuade us to abandon this assumption." Suppose, then, we were persuaded.

The nonidentity thesis would nevertheless have troubling implications for *death behaviors*, those behaviors we normally associate with someone's passing away. Assume that person essentialism is true and Al has passed away, leaving only a demented human animal. His loved ones – who understand person essentialism, let's say – know that Al is gone, because the existing patient is clearly not a person, even if no one knows when the transition occurred. Shouldn't they grieve in full force rather than waiting for the severely demented patient to die?[24] They should not treat the latter as if he were the individual they loved, because *ex hypothesi* he is not. He is a new individual. (We have noted the implausibility of regarding the severely demented man as having earlier coexisted with the person.) It would be very difficult, however, to convince loved ones to regard the severely demented man as someone new. While family members and friends sometimes grieve when a patient has become severely demented, their grieving is only partial. Rather than thinking that the individual is

---

[23] As explained in Chapter 2, I don't deny that constitution of one substance by another is a real relation – as, say, a hunk of bronze might constitute a statue and outlast the latter it if is melted. I deny that we are essentially persons and constituted by human animals.

[24] Cf. Olick, *Taking Advance Directives Seriously*, p. 143.

dead or entirely absent, they simply believe she is not (qualitatively) the person they knew and is no longer socially available in the same way. I suggest that the predictable reluctance to regard the severely demented man as a new individual reflects the insight that he is not. An alternative hypothesis is that reluctance to regard the severely demented man as someone new reflects the entrenched habit of associating the person who was with his living body, so my argument here is only suggestive.

Supposing loved ones *were* convinced that the demented man was a new individual, surely they wouldn't be ready for a funeral. And for good reason: The old man is still alive! Of course, one might insist that the person has passed away even if there is not yet a body to bury or cremate. But note how metaphysically odd this claim is: The man they loved passed out of existence *yet left no remains*. It would be odd to claim that his remains are precisely the body of the demented man, because the latter is still alive and surely owns his own body. Even if he is not a person, he is sentient and therefore has (at least some) moral status. A formerly existing person can't own an existing individual's body any more than I can own my living dog's body.

A champion of the nonidentity thesis might reply that, even if the formerly existing person can't straightforwardly own the body of the numerically distinct demented nonperson, the former has *surviving interests* in what happens to that body, just as one can have surviving interests in the disposal of one's later corpse in ordinary cases (where there is no issue about a surviving individual's identity). Thus Buchanan and Brock contend that "[t]he right of self-determination concerning what is to happen to one's living, nonperson successor could perhaps best be conceived as something like a property right in an external object, and as such more easily limited or overridden than the right of self-determination concerning what happens to one's self."[25] According to the argument, the present sort of case differs from ordinary cases of surviving interests in the disposition of a body in that here the body is "animated" by another living individual with potentially competing interests. That individual's interests or those of other parties, such as family members or society at large, could override the nonexistent person's surviving interests in others's honoring her advance directive. Perhaps this suffices to neutralize the charge that the nonidentity thesis implies that the person who passed away left no remains. (We will return to the topic of surviving interests. Of course, if the nonidentity thesis is false, then talk of surviving interests is

[25] *Deciding for Others*, p. 166.

out of place in the present context, because the original patient remains alive.)

The nonidentity thesis has further implications regarding death behaviors that will strike many as odd. For example, the prior person's spouse would now be a widow or widower, would be free to marry someone else (without first getting a divorce), and would presumably be entitled to life insurance.[26] Current practices, however, suggest that none of these death behaviors is appropriate until the demented nonperson has died. Although we could propose changing these practices on the basis of the nonidentity thesis, it is hard to believe that many people would favor such changes. While one could respond that such resistance would be irrational, I suggest, on the contrary, that it would reflect the commonsense insight that the author of the advance directive remains alive.

Finally, the nonidentity thesis apparently implies that the severely demented patient, as a new individual who has never earned an income or obtained health insurance, has no funds or insurance whatsoever. If so, why should the author of the advance directive, who supposedly no longer exists, have his funds posthumously drained? Or, if insurance covers the expenses, why should an insurance company reimburse care for someone they never insured? One might argue that the earlier person's advance directive serves, in effect, as having willed the use of his funds for the new individual – a case of surviving interests. Yet we standardly regard our advance directives as directing care for ourselves in the future. Those who claim that our standard assumption here is mistaken would still need to argue that nonidentity justifies taking the directive as a will *despite the fact that its author did not intend it as such*. Meanwhile, it is at least as difficult to understand why an insurance company should reimburse care for a new individual who is not covered under their plan. Perhaps one would argue that insurance companies would favor such a practice for public relations reasons, since so many people will deny the nonidentity thesis (irrationally, according to the argument). It seems more plausible, however, to maintain that continuity in coverage parallels the persistence of a single human individual.

In view of the fallacious reasoning underlying the nonidentity thesis, the implausibility of its premise – person essentialism – and the many incredible practical implications of the thesis, we may consider the Someone Else Problem (as it is normally construed) dissolved. The sentient,

---

[26] Cf. Olick, *Taking Advance Directives Seriously*, p. 143.

severely demented nonperson is not someone else from the author of the advance directive. They are numerically identical.

HAS NARRATIVE IDENTITY BEEN DISRUPTED?

One might accept the conclusion of the preceding section, yet claim that the authority of advance directives in cases of severe dementia is threatened by a different sort of Someone Else Problem. The problem, one might claim, is that while Al's numerical identity is maintained throughout the ravages of dementia, his *narrative* identity is disrupted. Al's sense of himself as the persisting protagonist of his self-narrative is now entirely absent. He is no longer psychologically connected to the life of the person who completed the directive. In this sense, although strictly speaking Al has persisted into his present state, he has in a very important sense become someone else from the earlier person. While the relevant sense of distinctness here is qualitative, not numerical, it captures the most profound sort of qualitative change possible in a human being: the termination of one's self-narrative. This disruption of narrative identity, the argument concludes, threatens the authority of the earlier advance directive just as if the earlier person and the present demented individual were numerically distinct. For all practical purposes, they might as well be distinct, because the present patient is psychologically cut off from his personal, narrative-carrying past.

This argument has considerable initial appeal. As discussed in Chapter 3, much of what matters to us in our existence over time is continuing our lives as persons with unfolding self-narratives and, if possible, successful self-creation. Yet Al-at-seventy-five lacks awareness of his earlier life in which he enjoyed such meaningful psychological richness. In his diminished mental condition, he does not identify, or connect himself narratively, with the earlier person. Disruption of the earlier Al's narrative identity, it may seem, is disruption of the only kind of identity that should count with advance directives.

Despite its initial appeal, this argument posing a narrative-based Someone Else Problem is problematic.[27] First, in claiming disruption

---

[27] Interestingly, authors who appeal to narrative in discussing advance directives consistently conclude that considerations of narrative do not threaten the authority of advance directives, although their supporting arguments differ significantly. See Mark Kuczewski, "Whose Will Is It Anyway? A Discussion of Advance Directives, Personal Identity, and Consensus in Medical Ethics," *Bioethics* 8 (1994): 27–48; Ben Rich, "The Values History: Restoring Narrative Identity to Long-Term Care," *Journal of Ethics, Law, and Aging* 2 (2)

of narrative identity, the argument views identity retrospectively from the standpoint of the demented individual rather than prospectively from the standpoint of the earlier person. As we will see, the prospective standpoint is no less important. Moreover, many people would identify with future stages of themselves in which they are severely demented even if they have difficulty anticipating clearly what such a state would be like. Indeed, individuals who have actively contemplated the possibility of becoming severely demented and have taken that possibility into account in completing an advance directive presumably not only identify, in some significant sense, with such future possible stages of themselves but care greatly about their fate during those stages.

Consider Al-at-sixty-five's prospective thinking. He suspects he is in the early stages of dementia. Fortunately, he still has enough reasoning ability, sense of his own priorities, and decision-making capacity to decide autonomously about his future care. Through his own research and conversations with Dr. Pedemonte, Al becomes well informed about the disabilities, remaining abilities, and other characteristics of patients with severe dementia, and he understands that demented patients are sometimes contented. Appreciating that in this condition he would no longer have a sense of his life as a whole, he nevertheless identifies with this likely future stage of himself.

But what does this mean exactly? He doesn't identify with the future Al in the sense in which people often identify with others they emulate or in the way one might identify with those of one's own characteristics one finds most agreeable. These two *evaluative* senses of "identify with" express a favorable evaluation of certain characteristics or persons embodying valued characteristics.[28] By contrast, Al finds the prospect of severe dementia ego-dystonic – disagreeable to his self-image. Nor does Al identify with his future self in the sense in which I once identified with an elderly neighbor: finding common cause with him due to shared values and interests, and believing myself similar to him in important ways.

(1996): 75–84 and "Prospective Autonomy and Critical Interests: A Narrative Defense of the Moral Authority of Advance Directives," *Cambridge Quarterly of Healthcare Ethics* 6 (1997): 138–47; Blustein, "Choosing for Others as Continuing a Life Story"; Michael Quante, "Precedent Autonomy and Personal Identity," *Kennedy Institute of Ethics Journal* 9 (1999): 365–81; and Jeffrey Spike, "Narrative Unity and the Unraveling of Personal Identity: Dialysis, Dementia, Stroke, and Advance Directives," *Journal of Clinical Ethics* 11 (2000): 367–72.
[28] So does the sense of identify employed in Chapter 3's analysis of autonomy, in which one can identify with one's own preferences.

This *empathic* type of identification involves having (or thinking one has) a good grasp of the other's perspective. It often overlaps with evaluative identification, but it need not: One might feel similar in certain respects to someone else, yet disapprove of the relevant characteristics and want desperately to become less like that person. Also inapplicable here is *fantasized identification*, which often occurs when we get caught up in a novel or movie. Here we empathize with a character, sympathize with her (not only understanding but caring about her perspective), and even, to some extent, feel as though we *are* that person. Thus, when terror strikes her heart, we are terrified; when her trust proves vindicated, we feel greatly relieved. Yet we know we are not that character, and our identification and associated intense feelings typically dissipate as soon as we return to our reality-based routine.[29]

Al's identification with his future self is reality-based and need not involve evaluation, because at its core is a sense of numerical identity. He identifies with the older Al in the sense that he believes, *and feels*, that he will in fact be that individual. And precisely because he accepts the fact of numerical identity between himself now and himself later, and finds the prospect of severe dementia so ego-dystonic, he chooses now to forgo life supports once he has entered that state.

We may clarify further possibilities for identification, and its connections with identity, by examining other examples. Suppose Franz not only accepts the fact that he may become severely demented – it would be *he* – but doesn't even find this prospect especially ego-dystonic. Franz values his life so much that he feels that continuing existence on this earth in any state is a great blessing: "While there would be some disadvantages to advanced senility, I can see myself making the most of it, enjoying interactions with others even if they don't seem familiar, welcoming whatever pleasures life brings even if I'm confused about where I've been and where I'm going." There would appear to be especially little basis for a narrative-based Someone Else Problem in Franz's case.

Now consider Angelica, who does not identify with a future severely demented self in either the numerical or the evaluative sense. Not only does she find the prospect of such a state highly disagreeable; she denies that it would be she who was in such a state: "Whoever that old woman would be, she wouldn't be me." But, if she means this literally, she is mistaken.

---

[29] For insightful remarks on empathy, sympathy, and identification, see Raymond Martin, *Self-Concern: An Experiential Approach to What Matters in Survival* (Cambridge: Cambridge University Press, 1998), pp. 96–9; also p. 127 on identificatory fantasies.

That is why she is generally regarded as having decision-making authority concerning medical care for the later severely demented individual.

Is this response to Angelica too literal, too focused on numerical identity? Imagine her replying as follows: "I get your point about numerical identity. What concerns me is that my sense of myself as a person with a particular life history will be gone if I'm severely demented. What's most important to me about my life – my memories, my relationships with family and friends, my most cherished values – will be mostly absent when I'm so senile. So I don't feel I'm really connected in any meaningful way to that possible future stage of myself." This understandable concern returns us squarely to narrative identity, this time from the prospective standpoint that's salient in completing an advance directive. We might say that while Angelica correctly *believes* she could become severely demented, she doesn't *feel* that that individual would be she. In this sense, she is alienated from and doesn't identify with that future stage of herself. By contrast, as we may put it, Al-at-sixty-five *robustly identifies* with Al-at-seventy-five: believing and feeling that the older man will be he.

Angelica could take her argument in either of two directions. She might argue that disruption of her narrative identity creates a Someone Else Problem such that she lacks authority to determine through an advance directive what will happen to the later severely demented patient. Alternatively, she might claim that numerical identity confers authority on the directive she is now completing ("After all, it would be me"), while the disruption of narrative identity ("I wouldn't be *myself*") justifies a decision now to forgo life supports at that later time. Both lines of argument assume that disruption of narrative identity is normatively important despite numerical identity. But in the second scenario, where Angelica invokes disruption of narrative identity to justify forgoing life supports, there is no special justificatory problem; she can invoke whatever prudential grounds seem salient in making decisions for her later demented self. Only the first scenario threatens to pose a significant Someone Else Problem. We will be better positioned to address this challenge, later in this chapter, after exploring a theoretical framework that arguably bolsters the case for such appeals to disrupted narrative identity.

VIEWS THAT VINDICATE ADVANCE DIRECTIVES ON THE BASIS OF PRECEDENT AUTONOMY

Having examined reasoning that allegedly generates a Someone Else Problem regarding advance directives, let us turn to views that reject such reasoning. These views either accept that the author of an advance directive and the severely demented individual to whom it apparently applies are numerically identical, or deny that they need to be in order for the advance directive to serve as a vehicle of precedent autonomy. Collectively, they represent the orthodox understanding that advance directives should play a major role in decision making for formerly competent patients on the basis of their earlier autonomous preferences. These views divide according to whether they accept the claim that numerical identity is necessary for the authority of advance directives.

Appeals to Precedent Autonomy on the Assumption That Numerical Identity Matters

According to the first cluster of views, numerical identity between, say, Al-at-sixty-five and Al-at-seventy-five along with the former's precedent autonomy are both necessary and jointly sufficient conditions for the (at least strongly presumptive) authority of his advance directive in medical decision making for Al-at-sevety-five. They also agree, without using quite this language, that precedent autonomy involves narrative identity.[30] Put schematically:

$$\text{Num ID} + \text{PA (Narr ID)} \to \text{authoritative AD}.$$

In order to clarify this thesis, we need to explain the concept of precedent autonomy, why numerical identity is necessary for the advance directive to carry presumptive authority, and how precedent autonomy connects with narrative identity.

To understand precedent autonomy requires clarity about autonomy more generally. In Chapter 3 we unpacked autonomous action as follows: *A autonomously performs intentional action X if and only if (1) A does X because she prefers to do X, (2) A has this preference because she (at least dispositionally) identifies with and prefers to have it, and (3) this identification has not resulted primarily from influences that A would, on careful reflection, consider alienating.* In the present context, intentional action *X* is completing an advance directive. As noted earlier, an advance directive is valid only if it was completed autonomously. Let us assume that this is so, taking Al's case as illustrating the satisfaction of the three conditions for autonomous action. First, Al completed the advance directive because he

---

[30] Quante comes close to this language ("Precedent Autonomy and Personal Identity," p. 366).

wanted or preferred to. Second, he would identify (evaluatively) with this preference if the issue were raised, inasmuch as it fits comfortably within his overall system of values and preferences. Third, (let's assume) such identification would not result from influences that Al would consider alienating.

Al completed his advance directive autonomously. But does the authority of his earlier decision extend to his present autonomy-lacking condition? How can it respect Al's autonomy to comply with a preference he no longer affirms or even understands? The question is especially urgent, because if the presently contented Al has any *current* preference about whether or not to continue living, it presumably is to live (though he is incapable of saying so explicitly). Or, if he has no preference about continuing to live because he is no longer capable of understanding the issue, one might argue that he nevertheless has an interest in continuing to live so long as he is generally contented.[31] So the idea of precedent autonomy is crucial.

The basic idea is that just as one's self-regarding preference, autonomously formed at time $t1$, can refer to how one is treated at a later time $t2$ when one lacks autonomy, others can respect or fail to respect one's autonomy at $t2$. Although one lacks autonomy at $t2$, the system of values and preferences that one had while autonomous extends to "objects" that survive loss of autonomy. While there can be conflicts between what one autonomously wanted regarding the current situation of incompetence (say, to forgo life supports) and what one now wants (say, to continue living), such conflicts should generally be settled in favor of one's *autonomous* preferences. Since the severely demented individual is presently incapable of autonomy, this reasoning vindicates a strong presumption in favor of honoring advance directives (even if other considerations, to be discussed later, sometimes justify overriding this presumption). Naturally, this assumes that there is a strong moral

---

[31] Helga Kuhse thinks otherwise, contending that sentient nonpersons such as higher non-human animals and the severely demented have no interest in continuing to live: "only persons can anticipate, and have desires about, their own future. These desires can be thwarted by a person being killed" ("Some Reflections on the Problem of Advance Directives, Personhood, and Personal Identity," p. 359). But the claim that no nonpersons have future-oriented desires is completely implausible. It implies that dogs and monkeys, for example, are incapable of fear, of any desires at all, and of even the simplest intentional actions. For a summary of arguments that support my contention, see my "Are We Essentially Persons? Olson, Baker, and a Reply," *Philosophical Forum* 33 (2002), pp. 113–14; for the arguments in full, see my *Taking Animals Seriously: Mental Life and Moral Status* (Cambridge: Cambridge University Press, 1996), pp. 129–83.

presumption in favor of respecting autonomy; later we will examine this thesis and its supporting arguments more closely.

For now, note that this reasoning also depends on an assumption of numerical identity. In the words of Ronald Dworkin, the most influential proponent of the present view, appeals to precedent autonomy must assume "that it is *correct* to regard a demented person . . . as a person who has become demented" – that is, "that the competent and demented stages of life are stages in a *single* life, that the competent and demented selves are parts of the *same* person."[32] (Dworkin uses the term *person* more broadly than we have, applying it even to the most severely demented individuals. This usage is common among lawyers, who tend to regard as persons all those, including demented patients, who have full protection of the law. To generate statements consistent with our usage, we could substitute *individual* for *person* in the preceding two quotations from Dworkin.) The essential idea is that one's autonomous self-regarding wishes can be respected even when one is no longer autonomous, so long as one still exists.

What about autonomous wishes concerning the distribution of one's assets and the disposal of one's corpse? Aren't these routinely respected despite one's nonexistence, since one must die before these wishes even become relevant? Yes, but the wishes in question aren't literally *self-regarding*; rather, they concern one's property and remains. For this reason, they lack the full force of autonomous self-regarding preferences. One's autonomous preferences regarding *oneself* have greater moral force than one's autonomous preferences regarding the posthumous handling of one's possessions (as suggested earlier in discussing surviving interests). So in order to justify a *strong* presumption favoring the authority of advance directives, appeals to precedent autonomy depend on the assumption of numerical identity.

Interestingly, precedent autonomy is closely connected to *narrative* identity. Your narrative identity involves your self-conception, your self-told story about your own life and what's most important to you. Rather than merely listing events involving you or facts about you, it orders these events and facts, highlights certain features and people, and organizes what's highlighted into a more or less coherent story of your life and character. One's significant decisions stem from the story as told up to

---

[32] "Autonomy and the Demented Self," *The Milbank Quarterly* 64 (suppl. 2) (1986), p. 4. He develops his view more thoroughly in *Life's Dominion: An Argument about Abortion, Euthanasia, and Individual Freedom* (New York: Knopf, 1993).

that point and project into the future in a way that expresses one's values and priorities. If one has self-regarding intentions that extend beyond the time when one has explicit narrative-telling capacity (roughly, robust self-awareness and decision-making capacity), one can autonomously make decisions for oneself in future times of nonautonomy. This is the idea of precedent autonomy. Extending the metaphor of narrative, the assumption is that one can write some of one's narrative in advance and include portions that will, or may, occur when one can no longer appreciate one's own narrative. Since a self-story is no mere recounting of facts as they occur, but an account that expresses one's values and priorities, which reach into the future, this extension of the metaphor is coherent and rather plausible, suggesting the coherence of the idea of precedent autonomy. (Showing that precedent autonomy has significant moral force will require further argument.)

This point motivates a distinction between two senses of narrative identity, which will later prove important. What we may call *strong narrative identity* or *persisting narrative capacity* is one's continued existence with significant psychological unity and the capacity for appreciating one's ongoing narrative. This is absent in the relationship between Al-at-sixty-five and Al-at-seventy-five, because the latter is so psychologically disconnected from the former. But, because Al-at-sixty-five makes self-regarding plans for Al-at-seventy-five, prospectively bringing the older man into his narrative, there is a sense in which the two are narratively identical. Let us call *weak narrative identity* the relation that is created when a person projects her self-narrative to a future time when she will exist without narrative capacity, as with decision making in accordance with precedent autonomy.

Different views that endorse precedent autonomy agree that certain preferences expressed prior to the onset of dementia (presumptively) trump preferences expressed by the severely demented individual. But how, more precisely, should we characterize the preferences that trump? Let me briefly outline three possible answers.

Dworkin's approach rests on a distinction between experiential and critical interests.[33] All sentient creatures have experiential interests, interests in activities or experiences the subject finds intrinsically pleasant

---

[33] See *Life's Dominion*, pp. 201–32. Rich also exemplifies this approach ("Prospective Autonomy and Critical Interests"). For an insightful critique of Dworkin's view, see Rebecca Dresser, "Dworkin on Dementia: Elegant Theory, Questionable Policy," *Hastings Center Report* 25 (November–December 1995): 32–8.

or enjoyable. Critical interests, by contrast, are aims, hopes, and projects whose attainment the subject regards as adding meaning and depth to her life – that is, significant prudential value – independently of whether she finds attaining these aims intrinsically enjoyable. Thus a person might have a critical interest in maintaining relationships with his siblings even if he usually doesn't enjoy their company or working to maintain the bonds. Another example is someone's interest in fulfilling her sense of moral obligation to contribute to her community, independently of whether she likes doing so. According to Dworkin, although both types of interests are important, critical interests take priority over experiential interests, contributing more to the integrity of one's life, to its narrative coherence. (Presumably, a more plausible claim is that someone's critical interests take priority over whatever she classifies as her noncritical interests; how she ranks particular experiential interests – as critical or noncritical – is left to her prudential judgment.) Now some demented patients who have experiential interests, whose attainment requires remaining alive, fail to remember their earlier-formed critical interest to avoid what they considered the indignity of massively compromised mental functioning. In such cases, according to Dworkin, the priority of critical interests requires that we respect precedent autonomy.

Another way to identify trumping preferences in accordance with the doctrine of precedent autonomy begins by distinguishing all-things-considered preferences, or those autonomously formed, and current preferences.[34] One may currently prefer *X* despite never having considered *X* versus non-*X* in a well-informed, careful state of mind that considers one's long-term values and priorities. If so, one's current preference is less likely than one's all-things-considered preferences to reflect one's long-term values and priorities. So when preferences of the two types conflict, there are strong grounds for overriding the current preference.

A third way to vindicate precedent autonomy and the associated preferences is to prioritize the ethical principle of respect for autonomy over that of beneficence. When beneficence toward an individual and respecting her autonomy conflict, contemporary moral thinking has generally favored the antipaternalistic approach of favoring respect for autonomy, allowing the individual to make self-regarding choices even where others doubt that a particular choice serves her best interests. Now one might

---

[34] See, e.g., John Davis, "The Concept of Precedent Autonomy," *Bioethics* 16 (2002), pp. 131–2; and Spike, "Narrative Unity and the Unravelling of Personal Identity."

deny that these two principles genuinely conflict, claiming that one's autonomous self-regarding decisions reflect what one believes is best for oneself; and since beneficence seeks the agent's best interests, and an autonomous agent is authoritative on what is in her best interests, beneficence and respect for autonomy converge.[35] As will become apparent in a later section, this convergence thesis regarding autonomy and best interests is probably oversimplified. In any case, there is broad, well-supported agreement that beneficence should generally give way to respect for autonomy when the latter genuinely applies to the same individual – that is, when the agent is capable of autonomous decision making. From this perspective, because the severely demented patient's recent preferences were not autonomously formed, respecting autonomy requires respecting precedent autonomy whenever the relevant, previously expressed preferences are known. (In a later section we will encounter the important objection that, where the psychological bonds between an individual at $t1$ and $t2$ are significantly attenuated, the moral importance of respect for precedent autonomy diminishes.)

Having sketched three arguments for allowing earlier autonomous preferences to trump current nonautonomous preferences, consistent with the doctrine of precedent autonomy, we need not adjudicate among these arguments. We may view them as friendly philosophical competitors among proponents of precedent autonomy who judge that numerical identity is a necessary condition for advance directives to carry authority. Let us now consider an issue that more significantly divides theorists in this camp.

How strong is the authority of precedent autonomy? There is agreement within this camp that precedent autonomy, conveyed through a valid advance directive, has strong presumptive authority such that it typically overrides any conflicting current preferences of the patient's. But what if a demented patient is *exceptionally* cheerful and physically comfortable? Many would argue that it is in the best interests of such an individual to keep living, setting up a clash between respect for (precedent) autonomy and beneficence. Others, who insist that best interests cannot conflict with precedent autonomy, admit that there are cases where it would be difficult not to provide life supports, despite an advance directive refusing them. These commentators might appeal to mercy for the patient – or even for the health professionals caring for him – as possible

---

35 Edmund Pellegrino and David Thomasma apparently support this thesis (*For the Patient's Good* [New York: Oxford University Press, 1988]).

grounds for overriding precedent autonomy in such cases.[36] Should such appeals to best interests or mercy ever prevail? Later we will find good reason to think so.

Appeals to Precedent Autonomy without the Assumption That Numerical Identity Matters

Another approach accepts the importance of precedent autonomy, but not of numerical identity, for the authority of advance directives. On this approach, precedent autonomy is intimately related to narrative identity, which is the type of personal identity that matters in this context; metaphysical considerations of numerical identity can be set aside. This view might be schematically stated this way:

$$PA \ (Narr \ ID) \rightarrow authoritative \ AD.$$

This approach can take a communitarian form. One argument for a communitarian approach to identity is that only such an approach accounts adequately for surviving interests. Earlier we rejected Buchanan and Brock's claim that the person who completed an advance directive has a surviving interest in how her severely demented "successor" is treated, because we rejected the assumption of nonidentity. A better candidate for a surviving interest is my interest in my daughter's flourishing. Assuming my daughter outlives me, my interest here has an object – her flourishing – that will extend temporally beyond my life. How can we make sense of an interest's persisting when the person who had the interest no longer exists? One possibility is to say that the interest survives in the broader community that supplies much of the content of one's self-narrative. As Mark Kuczewski puts it, "[such interests] survive in the memories, both personal and institutional, of the community. One's interest in his son's education survives because people remember his commitment to it; the arrangements the father made to provide for the education encode these intentions."[37] He applies this reasoning to advance directives, invoking the idea of narrative:

Likewise, an individual's interest in dignity, privacy, and bodily integrity is encoded in his living will, in the memory of others, and even in their perceptions of the situation at the time of one's incompetence. . . . It is a narrative sense of interest in that I conceive of myself objectively as part of a larger group that contains

---

[36] See, e.g., Dworkin, "Autonomy and the Demented Self," p. 13.
[37] "Whose Will Is It Anyway?", p. 42.

a "part" of me that transcends my individual consciousness and psychological continuity. . . . The body that belongs to the incompetent patient at t2 is in some sense "mine" because other persons will call it by my name and make the story of what happens to it a chapter of the story they tell about "me."[38]

These claims about advance directives fit well with our view of narrative identity. Whether such claims require a commitment to communitarianism – whose view of the self, some argue, is incoherent[39] – is another matter.

Like Kuczewski, Jeff Blustein argues that the salient sense of identity in the advance directives context is the narrative sense. But Blustein rejects communitarianism (for reasons we need not consider), developing his view instead on the basis of Marya Schechtman's narrative view (see Chapter 3). It will be instructive to consider Blustein's summary of his position:

(1) Persons cannot help but see themselves as temporally extended subjects ...and so have an interest in how their life stories will go and turn out; (2) Persons are justifiably concerned about a future when they can no longer narrate their lives, and that it should be in keeping with their self-definition; (3) The interest in the whole of one's life story, including those parts when one is permanently unconscious or severely demented, survives the loss of competence and narrative capacity; and (4) [Proxies] reflect this interest in the choices they make on the patient's behalf by making treatment decisions that cohere with the direction of the patient's life story, insofar as this can be ascertained.[40]

I agree with all four theses. What is indefensible, I suggest, is the further claim, also advanced by Kuczewski, that "we do not need to posit a continuing subject to make sense of surviving interests" and the authority of advance directives in cases of severe dementia or permanent unconsciousness – which suggests that, while narrative identity is necessary, numerical identity is not.[41] Let me explain why this is indefensible, beginning with surviving interests.

The commonly accepted notion of surviving interests – interests that survive after their subject goes out of existence – makes sense only if we don't take it very literally. It is true that people typically have some interests, usually rooted in specific preferences, whose objects will survive

---

[38] Ibid.
[39] See, e.g., David Luban, "The Self: Metaphysical Not Political," *Legal Theory* 1 (1995): 401–37.
[40] "Choosing for Others as Continuing a Life Story," p. 30.
[41] Ibid., p. 27. Elsewhere he suggests that numerical identity may be necessary after all (p. 26). Suffice it to say that Blustein's position on this issue is unclear.

them. Thus, I have interests in my daughter's flourishing after I die, in my wife's well-being if she survives me, and in the posthumous distribution of my asserts in accordance with my will. I have these interests *now*. Although in a sense they will survive me, they won't survive as *interests* in the usual or literal sense of the term – in which interests must have a subject. When I go out of existence (at the time of death – see Chapter 2), I have no interests or welfare of any kind. I can't have interests because I can't have properties. I can't have properties when I am dead (note the remarkably misleading turn of phrase "when I am dead") because at that time there will be no thing that is me. Properties are properties *of something*, and there will not be anything of the relevant kind to have my properties. My reputation, a phenomenon in the minds of other people, will exist and have properties, as will my remains, assets, some of my loved ones, and so on. But it is a grotesque error of philosophical logic to think I will have interests or any other properties when there is no me.

The somewhat loose sense in which there are interests that survive me is that there remain strong normative reasons for being responsive, after my death, to the "prospective posthumous" interests I had while alive. As virtue requires, we show respect to the person who was by honoring his will, keeping promises made to him, and not "disturbing" his corpse. Consistent practice of this respectful virtue is in everyone's interests because, while alive, we have preferences whose fulfillment or nonfulfillment will occur after we die. Such reasons for honoring surviving interests easily explain the importance we grant to the latter without illogically implying that a nonentity has interests. It also obviates the peculiar idea, advanced by champions of *backward causation*, that someone's action today (e.g., slandering the deceased) could retroactively affect the interests or well-being of someone who existed only in the past.

So Blustein is correct that we need not posit a persisting individual to make sense of surviving interests. But, as a literal reading of his own words suggests, we do have to posit a persisting individual to make sense of the authority of advance directives: "The interest in the whole of one's life story, including those *parts when one is permanently unconscious or severely demented*, survives the loss of competence and narrative capacity" (emphasis mine). As stressed earlier, if the advance directive's author is not the individual to whom it would apply, it lacks authority. If I were essentially a person, and the future severely demented nonperson merely my "successor," I would have no autonomy-based prerogative to determine his fate. While the authors are right that a continuing narrative is crucial to the authority of an advance directive, they are mistaken in denying that numerical identity is also necessary.

SKEPTICISM ABOUT PRECEDENT AUTONOMY

The previous two sections examined a pair of approaches that concur that precedent autonomy establishes the presumptive authority of advance directives. Some commentators, however, are skeptical about appeals to precedent autonomy. They argue that the grounds for skepticism either vitiate the presumptive authority of advance directives or eliminate this authority altogether.

In cases of severe dementia, where the demented patient is not a person (in the sense discussed earlier), one fundamental reason for denying the authority of an advance directive is the nonidentity thesis, the claim that the author of the directive and the current demented patient are numerically distinct. We have undermined this thesis, which is based on person essentialism and the claim that the demented patient is a non-person (or, somewhat less plausibly, a *new and different* person[42]). The skeptics' stronger arguments stress qualitative differences in an individual at two different times.

Progressive dementia typically produces massive psychological changes. The changes in Al, for example, are not unusual. So one specific reason for skepticism about precedent autonomy is the difficulty of knowing, while competent, what life will or may be like as an individual whose mental life is very different from what it used to be. Note that we regard the consent of a competent patient as informed consent only if the patient has a reasonably strong understanding of her medical condition, the likely consequences of relevant options, and so on. Similarly, in order for someone to be sufficiently informed for her advance directive to count as valid, according to Dresser, we would want her "to understand

---

[42] See Dresser and Robertson, "Quality of Life and Non-Treatment Decisions for Incompetent Patients," p. 236. The assertion of a new person is implausible because, if the currently demented individual is a person at all – retaining the capacity for complex forms of consciousness – then surely she retains nontrivial temporal self-awareness or psychological continuity with the past. Hence one and the same person over time. If, however, this inference is denied on the ground that here the term *person* is being used more broadly to include individuals who only formerly had such self-awareness but remains sentient, then the retention of sentience undermines the claim of numerical distinctness. The person or individual has changed a lot qualitatively, but she has retained a property that allows one to continue to qualify as a person (as this argument construes personhood).

that the experience of dementia differs among individuals, that for some it appears to be a persistently frightening and unhappy existence, but that most people with dementia do not exhibit the distress and misery we competent people tend to associate with the condition."[43] We also have to allow for changes in heart. Someone may complete a directive with a particular set of intentions, later (while still competent) change his mind, yet never modify or replace the old directive. If we have convincing testimony from family members or other persuasive evidence that he changed his mind while competent, there is a strong case for overriding the advance directive, which no longer appears to reflect the patient's most up-to-date autonomous wishes.[44]

While concerns about sufficient information and possible changes of heart are certainly legitimate, they are more plausible as grounds for caution than as arguments against the authority of advance directives. Somewhat similarly, concerns about information and changes of heart in the context of informed consent motivate disclosure requirements and, in some contexts such as participation in research, an understanding that individuals may withdraw consent they had earlier provided. But these concerns do not vitiate the moral authority of informed consent. In the advance planning context, health professionals should do what they reasonably can to promote education about dementia for those completing advance directives; and evidence about possible changes of heart should be shared with the health care team. But, if the case for respecting precedent autonomy remains standing, the proper role of these concerns is to highlight possible grounds for rebutting the presumptive authority of directives. Accordingly, if strong evidence emerges that a demented patient was misinformed or uninformed about dementia *in such a way*

---

[43] "Dworkin on Dementia," p. 34.
[44] Why favor the most recent autonomous wishes rather than, say, those autonomously held for the longest portion of one's life? Suppose Lori, a liberal agnostic, for decades wanted not to receive life supports in the event of irreversible incompetence. After a religious conversion, she autonomously changed her mind and wanted life supports. (Assume that both the earlier and more recent preferences are documented adequately.) Why prioritize her most recent autonomous preferences? If we counted her years as a competent adult equally, we should favor the earlier, longer-held preferences. Perhaps the best reason to favor the most recent autonomous preferences, as Robert Wachbroit reminded me, is that only the most recent autonomous self enjoyed the full perspective of all of the previous experiences and value judgments of the person's life. That is, having undergone her recent religious conversion, Lori is able to take into account whatever experiences were salient in her conversion as well as all prior experiences that shaped her agnostic, liberal outlook. To prioritize her preconversion values would be to neglect an important part of her history.

that would apparently affect the content of her advance directive, this evidence would cast serious doubt on the authority of the directive. Alternatively, if strong evidence emerges that a patient had a change of heart *such that his most recent autonomous preferences are contradicted by his advance directive,* this evidence would vitiate the authority of the directive. But these concerns do not appear to undermine precedent autonomy as a basis for the *presumptive* authority of advance directives.

A more significant challenge to precedent autonomy in the advance directives context adduces the significant psychological discontinuity between, say, Al-at-sixty-five and Al-at-seventy-five (without denying numerical identity). According to this challenge, such massive psychological change makes it irresponsible to assume a convergence between (1) the patient's past autonomous wishes and (2) her current best interests. Even if we cannot sharply distinguish the patient's former interests and current interests based on a claim of numerical distinctness, in view of this chapter's challenge to that claim, we can underscore the distinction based on the current patient's inability to remember, much less care about, what she wanted when competent. Being in this way psychologically separated from her earlier self, the severely demented patient has independent best interests based on her current needs and preferences. And the patient's current best interests should take priority.[45]

This important challenge to precedent autonomy begs two key questions. First, it assumes that the demented patient's best interests are a function of what she can now appreciate. As we have seen, Dworkin disagrees, holding that one's critical interests – which in the case of severely demented patients may be expressed only through precedent autonomy – trump experiential interests, which such patients will continue to have. A more general thesis that includes, without being limited to, Dworkin's assertion is that an individual's best interests are to be understood from a longitudinal or whole-lifetime perspective rather than from a perspective rooted in the individual's present understandings, preferences, and value judgments (if any).[46]

---

[45] See Dresser and Robertson, "Quality of Life and Non-Treatment Decisions for Incompetent Patients."

[46] Agnieszka Jaworska helpfully notes that many demented patients are not yet so cognitively impaired that they have lost their capacity to value. More controversially, she argues that to the extent that a patient retains this capacity, respecting her current best interests is compatible with respecting her autonomy ("Respecting the Margins of Agency: Alzheimer's Patients and the Capacity to Value," *Philosophy and Public Affairs* 28 [2000]: 105–38; cf. Howard Klepper and Mary Rorty, "Personal Identity, Advance Directives, and Genetic Testing for Alzheimer Disease," *Genetic Testing* 3 [1999]: 99–106). I disagree insofar as I hold that the capacity to value is not sufficient for autonomy (see Chapter 3).

The present challenge also assumes, contrary to the doctrine of precedent autonomy, that the demented patient's best interests, however determined, take priority over respect for the patient's precedent autonomy. Perhaps the most plausible way to defend this assumption is to argue that the moral authority of precedent autonomy decreases in proportion to the attenuation of psychological connections over time – and that, at least in cases of severe dementia, such connections are so attenuated that the best interests standard becomes more important than respect for precedent autonomy. This claim dovetails with another discussed earlier: that the radical disruption of the original patient's (strong) narrative identity creates a sort of Someone Else Problem regarding that patient once she has become severely demented. We may combine the two claims into the thesis that, *even if numerical identity obtains between the earlier and later selves, the disruption of (strong) narrative identity between them renders the later self's best interests independent of, and more important than, the earlier self's preferences and interests.*

We will evaluate the two assumptions underlying the present challenge to precedent autonomy after examining in the next section a novel theoretical framework for understanding best interests.

THE IMPORTANCE OF TIME-RELATIVE INTERESTS

In addressing the issue of rational egoistic concern, or what matters in survival (see Chapters 2 and 3), Jeff McMahan develops an account that will prove useful to the present discussion. This account refers to a being's *degree of psychological unity over time*, which involves three factors: "the proportion of the mental life that is sustained over that period, the richness or density of that mental life, and the degree of internal reference among the various earlier and later states [example of strong internal reference: a detailed memory of an earlier experience]."[47] Rabbits, for example, have considerably less psychological unity over time than normal humans have. So does Al between the ages of sixty-five and seventy-five. McMahan advances this thesis about the basis for rational egoistic concern:

The relation that is [on his view] constitutive of identity – sufficient physical and functional continuity of the areas of the brain in which consciousness is realized in order for those areas to retain the capacity to support consciousness – is both a necessary and sufficient condition of a *minimal* degree of rational egoistic concern. Beyond that, the *degree* of egoistic concern that it is rational to have about the

---

[47] *The Ethics of Killing: Problems at the Margins of Life* (Oxford: Oxford University Press, 2002), p. 75.

future may vary with the degree of physical, functional, and organizational continuity in [the relevant brain parts, corresponding to] the degree of psychological
unity within the life.[48]

On this view, the extent to which you should be egoistically concerned
about some possible event in your future – say, retiring with the mortgage
paid off – varies not only with the positive or negative value of that event,
as we ordinarily assume, but also with the degree of psychological unity
between yourself now and yourself later.[49] Hence a discount rate for
diminished psychological unity. This discounting determines the strength
of one's *time-relative interests* regarding possible events in your future, such
as your present interest in retiring with the mortgage paid off. If identity
were the sole basis for rational egoistic concern, then one's time-relative
interests would simply be one's interests – which are assessed from the
standpoint one's life as a whole, not relativized to some particular time.[50]
Where psychological unity is greatly diminished, as with a newborn baby
in relation to herself in the future, one's time-relative interests can diverge
sharply from one's interests – and prudential evaluation, on this view,
should rest on the former.

The Time-Relative Interests Account (TRIA) has important implications for the harm of death in the cases of animals and human nonpersons, and therefore for the ethics of killing animals and (as discussed
in Chapter 7) the abortion issue. It also has important implications for
decision making by competent individuals regarding their own futures
of possible incompetence and proxies' decision making on behalf of the
formerly competent. These latter two sets of implications are our present
concern.

Before turning to those implications, let us consider the case for the
TRIA. Why believe that psychological unity affects the degree of rational
egoistic concern about a future event or period of time, setting up a contrast between one's interests and one's time-relative interests? In short,
this thesis best explains certain prudential judgments – for instance, that
the death of a fetus who has attained the capacity for consciousness and
had bright prospects in life is not prudentially as tragic as the death of a

---

[48] Ibid., p. 79 (emphases added).

[49] On McMahan's view, another "prudential unity relation" (relation grounding rational
egoistic concern), besides degree of psychological unity, is identity itself, which on his
view consists in the persistence of a single embodied mind. Our discussion will focus on
psychological unity because this relation, unlike identity, admits of degrees and therefore
proves crucial to the comparative claims vindicated by this account.

[50] Ibid., p. 80.

fifteen-year-old child is; that a barely sentient creature with little psychological unity over time has little stake in continuing to exist (as opposed to having a good quality of life while he does exist); and that it *may* be rational for you to choose five more years of normal life over many more years of higher-quality life in which you would be psychologically discontinuous from yourself now.[51]

The essential idea is that prudential evaluation of a possible future – whether a single event, a specific period of time, or one's entire future – should take into account both the value of that future to one as one experiences it and how psychologically "invested" one now is in that future. It is reasonable to suppose that abortion ordinarily deprives a fetus of something valuable: a future of value to it.[52] Or, if one holds that only sentient beings can have interests, one might assert only that sentient fetuses lose a future of value. Yet it plausible to think that a toddler, for example, loses more than the sentient fetus does, because the toddler has developed a sense of herself, some future-oriented desires including desires involving emotionally developed relationships, and the like. I find it plausible that a twenty-year-old who has developed some life plans and has invested considerable energy into becoming a certain sort of person loses still more, typically, from premature death. (In making these comparisons, I am not drawing inferences about the ethics of killing human individuals at different stages of life.) Notice, however, that if we considered the harm of death solely from a whole-lifetime perspective, it would seem that the twenty-year-old, who has already lived quite a few years and has a shorter life expectancy, would lose the least from dying, and the two-year-old would lose somewhat less than the fetus. Yet precisely the opposite seems true. I think McMahan's proposal for explaining such judgments is on the right track: We must take psychological unity across time into account, so that the harm of death – or, more generally, the (dis)value of some possible future – is understood in terms of time-relative interests.

Further intuitive support for the TRIA comes from cross-species comparisons of the harm of death. Compare an ordinary person's death and that of a turtle. As vertebrates, turtles are surely sentient.[53] Their lives include certain pleasures (as well as pains) and perhaps other sources of value connected with species-typical functioning. Ordinarily, a turtle's

---

[51] Ibid., pp. 75–8.
[52] Chapter 7's discussion of abortion will examine this argument closely.
[53] Elsewhere I have argued that empirical evidence strongly supports the thesis that vertebrate animals are sentient (*Taking Animals Seriously*, ch. 5).

death involves a loss, the loss of opportunities for whatever goods his life would contain if he survived. But surely a turtle has very little self-awareness over time – in my view, enough to perform intentional actions but not enough for life plans or well-developed narratives.[54] If one turtle were to die before living out his natural life span and another with equal prospects for a good turtle life came into existence, that would not seem significantly worse, if worse at all, than for the first turtle to survive without another coming into existence. Persons, by contrast, do not seem so nearly replaceable. If Joe, who has a decent life, dies (painlessly, say) in middle age, it would hardly compensate to bring someone else in existence, because Joe's well-developed self-narrative would be cut off. With a turtle, it's almost as if the future he loses due to death might as well have belonged to another turtle.

A highly plausible explanation for this judgment is the TRIA. When the psychological unity that would have bound an individual at the time of death to himself in the future, had he lived, is weak, death matters less prudentially – that is, for that individual – at that time. This suggests that the turtle's death does not prudentially matter very much to him, when he dies, because the turtle's mental life is psychologically not very unified over time. I hereafter assume the correctness of this account (and will discuss it further in Chapter 7), although I will partly disagree with McMahan on how to apply it.[55]

The TRIA can be combined with Chapter 3's account of what matters. Addressing the question "How concerned should we be with an event or period of time in our future?" McMahan answers with a discount rate based on the degree of psychological unity between oneself now and oneself later. Chapter 3's reflections on narrative identity and self-creation speak, at a very general level, to the *content* of what typically matters to us in survival. The two views are broadly compatible and mutually reinforcing. A high degree of psychological unity over time guarantees the persistence of a person with an ongoing self-narrative (assuming numerical identity is maintained); and to the extent that the agent fulfills earlier-formed aspirations, contributing to self-creation, this deepens psychological unity.

---

54 Ibid., chs. 6 and 7.

55 Dan Brock asked me whether I can coherently reject McMahan's embodied mind account while embracing his TRIA. Yes, because the two accounts are logically independent. Chapter 2 contains my critique of the embodied mind account. Chapter 3 argues that, while wrong about numerical identity, psychological theories help to illuminate narrative identity and what matters in survival. The present discussion integrates the TRIA into the framework for understanding narrative identity developed in Chapter 3.

Why is the TRIA important to our discussion? Earlier we left open whether we should understand the severely demented patient's best interests as a function of (1) her present understandings and preferences or (2) a whole-lifetime perspective. Dworkin's view implies that we should understand Al-at-seventy-five's best interests from a whole-lifetime perspective, in which case his best interests are determined by his critical interest, expressed through precedent autonomy, to be allowed to die. Note that this approach assumes that numerical identity is very important to what matters in survival and that the disruption of (strong) narrative identity poses no threat to the authority of precedent autonomy. But the discussion of this section has distinguished one's interests, as understood from a whole-lifetime perspective, and one's time-relative interests, which are relativized to a particular point in time. Perhaps the latter provides a better account of Al-at-seventy-five's best interests.

I find this suggestion from McMahan to be intuitively compelling:

> [W]hen the prudential unity relations [the relations grounding rational egoistic concern] are weak between an individual now and herself during most of the rest of her life, what is good for her now is determined less by the way in which her present life contributes to the value of her life as a whole and more by her nature and preferences at the time.... This is because the less prudential unity there is within a life, the less the life matters *as a whole*.[56]

The less psychologically unified a life is, the more quality of life at a particular time matters and the less the life as a unit matters in determining someone's best interests. For example, your cat's best interests would seem to have much more to do with quality of life at a particular time than with some feature involving the cat's life as a whole. Supposing you had to give your cat away, and in her new living situation running in a nearby woods would improve her quality of life, it would hardly matter that this recreational activity did not "cohere" with her previous apartment-bound existence.

McMahan's suggestion may appear to recommend construing Al-at-seventy-five's best interests in terms of his *present* time-relative interests rather than his interests from a whole-lifetime perspective. But this doesn't follow. Even if prudential evaluation should track one's time-relative interests wherever they and one's interests (which are understood from a whole-lifetime perspective) diverge, we cannot forget that Al-at-sixty-five also had time-relative interests – which coincided with

---

[56] *The Ethics of Killing*, p. 500.

his interests. When he completed the advance directive, Al's most fundamental (time-relative) interest was to live a life that was true to his self-conception, a life not spoiled by a protracted period of what he considered degraded existence toward the end. Now that he is severely demented and psychologically cut off from the self-narrative that generated his interest in avoiding severe dementia, he has independent time-relative interests. Given his state of contentment and his cognitive dissociation from his earlier priorities, it would seem to be in his current time-relative interest to continue living. From this perspective, rooted in the present, the whole-lifetime perspective does not matter much.

Between the time-relative interests of Al-at-sixty-five and those of Al-at-seventy-five, which should take priority in determining his best interests? It might seem that we should favor his current time-relative interests because our question concerns the present: What should Dr. Pedemonte do now? But the present tense of this *question* does not entail that consideration of Al's best interests must favor the present over his entire lifetime. We still have the question of how to determine his best interests.

On the one hand, perhaps we should favor his earlier interest in a lifetime of intellectual vigor, because Al's current time-relative interest in remaining alive is relatively weak due to the weak psychological unity between him now and him in the near future; and he could not live very long in any event. Although he now has a time-relative interest in staying alive, it is a much weaker interest than Al-at-sixty-five's time-relative interest in dying should he become severely demented. On the other hand, the weak psychological connections pointing backward from present to past seem to neutralize this argument. I suggest that we have a stalemate: Considerations for the two opposing ways of construing Al's best interests are about equally reasonable and weighty, canceling each other out. There is, I suggest, no uniquely correct answer to the question of whether continuing to live would be in Al's best interests.

But this indeterminacy concerns only best interests. Precedent autonomy has independent moral importance in connection with the principle of respect for autonomy. Our study of precedent autonomy vindicated presumptive authority for advance directives. This presumption is rebuttable in certain types of cases, as we have seen. When it is rightly overturned, we should attempt to make a well-grounded substituted judgment. In some cases – for example, when we know that the patient had a change of heart while still competent, yet did not revise her advance directive – we can meaningfully apply the substituted judgment standard. In other cases – say, when the patient's lacking certain information was likely material in his decision making – the grounds for rebutting the advance directive are also grounds for rebutting a substituted judgment. In these cases, we should apply the best interests standard. But the present case does not seem to warrant overriding the advance directive's authority. (Even if it did, the best interests standard does not provide guidance.)

Suppose, however, that matters were otherwise and there were a convincing case that Al's best interests favored continuing life. This would set up a conflict between respect for autonomy and beneficence. Nevertheless, we must remember that we generally favor respect for autonomy when that principle and beneficence apply to the same individual. To think otherwise would require making one of two claims. One possibility would be to reject this antipaternalistic priority of respect for autonomy. To the contrary, with virtually everyone in contemporary bioethics, I accept the broadly antipaternalistic approach. Another possibility would be to accept antipaternalism but deny its relevance here. One might claim, with Dresser and Robertson, that precedent autonomy is mythical in cases of severe dementia, so the principle of respect for autonomy simply doesn't apply in these cases – in which case beneficence wins by default. We have rejected this option as well, vindicating a presumption favoring precedent autonomy while acknowledging the appropriateness of overriding that presumption in certain cases.

The case of Al is not one of those cases. He was well informed about dementia and the implications of his advance directive. There is no reason to think that, had he known even more about dementia, he would have decided differently. *Importantly, he made his decision in the understanding that he might be contented in his state of dementia.* Moreover, he had no change of heart before losing competence. Even if it were in his current best interests to continue to live, his precedent autonomy makes his life as a whole a very significant consideration (though not unambiguously a best-interests consideration) – and the best option from his whole-lifetime, autonomy-infused perspective, I submit, is to be allowed to die in circumstances like the present ones.

Admittedly, this is a hard case. It is especially difficult when we consider a powerful rejoinder that was noted *en passant* in earlier sections. The rejoinder is that, even if the attenuated psychological connections between the earlier person and the later severely demented patient do not eliminate the authority of precedent autonomy, they diminish its authority. We might say that disruption of narrative identity makes the severely demented patient "partly" someone else from the earlier person.

In that case, just as decreased psychological unity justifies a discount rate for prudential concern (at least on the TRIA), perhaps it should justify a discount rate for decision making authority as well.[57]

Before replying to this challenge, I must concede that were the case slightly different and Al had *not* evinced his understanding that he might be contented in a state of dementia, I would reach a different judgment. There would be genuine doubt about his preference to die in the present circumstance. Because a prospective preference to die in a certain medical condition *even in the event that one is contented* is highly unusual, a reasonably confident judgment that respect for autonomy involves allowing one to die requires that the patient explicitly articulate this preference. Had Al not articulated what he wanted in the event of pleasant dementia, it would be quite unclear how to respect his autonomy, leaving best interests as the appropriate guide. And, because his best interests from a whole-lifetime perspective would be uncertain for the same reason that respect for autonomy is, Al's best interests would be a function of his present time-relative interest, which is to live.

In the case as it stands, Al's earlier preference regarding his present circumstance is clear, so respect for autonomy recommends letting him die. Our challenge is the claim that respect for autonomy should carry less weight due to diminished narrative unity, in which case this principle would compete with best interests without either principle enjoying an obvious presumption in its favor. I believe this challenge does not carry the day for two reasons. First, even if the challenge is correct and respect for autonomy deserves reduced moral weight, it still deserves some moral weight and would prevail given the indeterminacy of best interests. Second, it is not clear that this challenge succeeds. My tentative, uncertain suggestion is that diminishing the authority of precedent autonomy makes sense only if the earlier person did not robustly identify with the later self. Al-at-sixty-five robustly identified with his later, demented stage of himself, believing and feeling that he would be that individual. I suggest that, when someone's prospective autonomous wishes are clear for a later circumstance in which he finds himself, they apply with full authority if the individual robustly identifies with the later self. In effect, the autonomous agent decides how much certain possible changes, such as those of dementia, matter to him in terms of his relationship to himself over time.

---

[57] Thanks to Maggie Little and Marya Schechtman for bringing this challenge to my attention.

If my tentative response to the present challenge is incorrect, then the advance directive should have diminished weight. But, once again, the indeterminacy of Al's best interests would favor the advance directive as a source of guidance. Thus, in the case that opened this chapter, Dr. Pedemonte should make Al-at-seventy-five comfortable, but withhold antibiotics and allow him to die.

CONCLUDING REFLECTIONS

As we have seen, personal identity theory helps to illuminate the authority of advance directives in cases of severe dementia. Investigating the concrete problem of the authority of advance directives, in turn, has motivated refinements of our accounts of what matters and of narrative identity. Here it will be helpful to recapitulate, or in some cases state more explicitly, some major theses that our discussion has supported. This will lead us to an issue, not yet discussed in this chapter, which we will also address.

Several Theses

The first thesis is metaphysical, while the next few concern decision making for incompetent patients. (1) *Not even the most severe dementia entails the death or passing away of the individual who was a person.* Even though Al-at-seventy-five is not a person, he is numerically the same individual as the person Al-at-sixty-five. So there is no Someone Else Problem, construed in terms of numerical identity, for the authority of advance directives. (2) *Just as an agent can have autonomous preferences regarding periods of her life when she is not autonomous, respect for autonomy can be a relevant principle during such periods of nonautonomy or incompetence.* This is the idea of precedent autonomy. (3) *The authority of advance directives, even when they are valid and apply in the case at hand, is rebuttable.* Evidence for a change of heart, for example, might overturn the authority of a valid advance directive. (4) *A severely demented patient's best interests may be understood from either a whole-lifetime perspective or as a function of her current time-relative interests; if these two interpretations conflict in a given case, the patient's best interests are ambiguous.* Dworkin's claim that best interests are determined by critical interests, as expressed through precedent autonomy, proved oversimplified. There are equally good reasons to understand best interests in terms of the demented patient's current time-relative interests. (5) *If a patient's best interests are indeterminate, while her prior preferences are clear and were clearly autonomous (e.g., sufficiently well informed), the earlier preferences should govern decision making.* This conclusion follows from the claim that precedent autonomy has at least some authority – a claim compatible with the thesis that the authority of advance directives decreases when a patient has little psychological unity over time. (I also tentatively suggested that this thesis about diminished authority is correct, if at all, only when a patient did not identify robustly with her later self.)

Several further theses address the basis for egoistic concern at a general level. (6) *Numerical identity is a necessary condition for rational prudential concern.*[58] If Al-at-seventy-five were numerically distinct from Al-at-sixty-five, then the younger man's preferences would not have the status of precedent autonomy in relation to the older man's treatment. And while we can have interests in the posthumous disposal of our remains and assets, we have these interests only so long as we exist – as living human individuals.

While numerical identity is necessary, I do not claim that it is sufficient. In one sense it is clearly not sufficient for what matters, at least to most people. For few would prudentially value a future of permanent coma. In another sense, it is harder to say whether numerical identity might be sufficient. While few of us would value being in a permanent coma, many of us would *disvalue* this state, implying that we have an interest in avoiding indignity or any state that offends our sense of what's valuable in our lives. As our discussion of surviving interests clarified, however, some of the interests we have as persons concern times when we do not exist and therefore when we have no interests. Does the permanently comatose individual really have an interest in avoiding indignity or is this a simply an interest of the former person, an interest that isn't literally ascribable to the patient now? This seems a matter of reasonable disagreement, and I will not attempt to settle the issue. The following, however, is clear: (7) *Assuming numerical identity, strong narrative identity (persisting narrative capacity) is sufficient for rational prudential concern.*

The concept of weak narrative identity also emerged in our discussion. As the doctrine of precedent autonomy implies, a person can project her self-narrative into a future in which she no longer has narrative capacity. Thus, Al-at-sixty-five is weakly narratively identical with Al-at-seventy-five. Thus I suggest: (8) *Assuming numerical identity, weak narrative identity (at*

---

[58] At least in all cases we are likely to encounter. Cases involving fission or cerebrum transplants, as discussed in Chapter 2, may constitute exceptions.

least when the individual retains the capacity for consciousness) is necessary and sufficient for rational prudential concern.

But suppose Al-at-seventy-five is provided antibiotics and continues to live. Now Al-at-*seventy-six* enters a permanent coma. Numerical identity, of course, is maintained. But are Al-at-sixty-five and comatose Al-at-seventy-six even weakly narratively identical (assuming the younger man had treatment preferences in the event of coma)? This is highly debatable.

On the one hand, Al still exists in a coma, and his earlier preferences included preferences about what to do with him in this state. One might argue that precedent autonomy extends this far – a claim consistent with the formula Num ID + PA (Narr ID) → authoritative AD and our legal convention of permitting advance directives to direct one's care in PVS or permanent coma. On the other hand, perhaps a living individual who has become permanently unconscious no longer has interests, so that the preferences expressed previously regarding what to do with him in this state are really only surviving "interests" – normative considerations that continue to have force, not interests the comatose individual presently has. Again, I remain agnostic on whether patients in PVS or permanent coma have interests. Even if they do not, though, one might argue that precedent autonomy can extend to states of permanent unconsciousness *so long as one is still alive and therefore still exists.* Whether because living, permanently unconscious former persons can have interests, or because precedent autonomy can extend to times when one lacks interests so long as one still exists, I am inclined to accept the claim that precedent autonomy and weak narrative identity can extend to PVS or permanent coma. If that is correct, we may revise thesis (8) by deleting the parenthetical qualification.

We have been discussing whether it is rational to have prudential concern for a future stage of oneself. In cases where it is rational, a distinct issue is *how much* prudential concern it is rational to have. McMahan argues that, at any given time, our time-relative interests determine how concerned we ought to be about something in our future. This has an interesting implication for advance directives. In discussing their authority, we have focused on decision making for the presently incompetent patient. Let's now consider decision making by the person completing the advance directive. How much should Al-at-sixty-five care about what happens to future stages of himself in which he is deeply demented?

Al's mental life will not feature very much psychological unity over those ten or so years. Why does this matter? McMahan, remember, defends a discount rate for diminished psychological unity, and this discounting determines the strength of one's time-relative interests. Nor, on his view, is this discounting merely permissible: "[P]rogressive dementia is, in a rough way, a mirror-image of our earlier psychological development.... But if we are rationally required to discount the fetus's or infant's time-relative interests in its own future because of [diminished psychological unity], it seems that we should maintain this insistence in the case of the [patient looking ahead at dementia] as well...."[59] But Al-at-sixty-five cares *deeply* about what will happen to him in a state of severe dementia. On McMahan's view, the intensity of his prudential concern is irrational, because his current time-relative interest in what happens in that later scenario is relatively weak.

This is where I part ways with McMahan's understanding of the TRIA. Al-at-sixty-five robustly identifies with the later demented stage of himself. Not in the sense that implies positive evaluation – again, he regards the prospect of dementia as ego-dystonic – but in the sense that he believes and feels he will be that individual and cares about his fate. Could he reasonably decide that the weak psychological bonds between him now and him later justify caring less, prudentially, about Al-at-seventy-five's fate than he cares about the fate of Al-at-sixty-six (with whom he is more strongly psychologically connected)? Yes, this is a reasonable option. Must he, on pain of irrationality, apply a discount rate in this way?

There is a crucial asymmetry, overlooked by McMahan, between Al as a fetus or infant and Al-at-sixty-five. The latter is an autonomous decision maker, the author of his self-narrative. He is numerically identical to Al-at-seventy-five. But it's up to him, I suggest, to determine *how much* he identifies with a later deeply demented stage of himself and, therefore, up to him to decide how much he cares what happens to him then.[60] My thesis is not simply that he is in a position to decide that he identifies strongly and cares deeply (egoistically) about his future demented self – an obvious point – but that there is nothing irrational in his doing so.

This thesis is connected with the observation that, unlike numerical identity, narrative identity implicates an agent's values. As discussed in Chapter 3, who you are (narratively) has much to do with what you care about and how you see yourself. Now it's a fact that Al-at-sixty-five is numerically identical with Al-at-seventy-five. It's also a fact that there is weak psychological unity between the two Al stages. But to treat it as a further

[59] *The Ethics of Killing*, p. 496.
[60] Although numerical identity is an all-or-nothing relation – *A* either is or is not one and the same entity as *B* – the psychological phenomenon of identification admits of degrees.

fact that Al-at-sixty-five *should* discount the prudential importance of what happens to Al-at-seventy-five would be to intrude, paternalistically and excessively, into the younger man's value judgments and sense of self. If this is correct, the TRIA should become *optional* in the case of autonomous persons thinking prudentially and prospectively. Thus, while Angelica from our earlier case cannot reasonably decide that her future deeply demented self is entirely distinct from her, and in that sense someone else, she may (rationally) discount her degree of prudential concern about the later self in proportion to the diminished psychological unity between her present and future selves. Such discounting might even lead her to decide not to complete an advance directive.

Thus: (9) *In the case of autonomous persons thinking prudentially about their futures, it is rationally permissible but not mandatory for them to discount – in proportion to their decreased psychological unity over time – the importance of future events affecting them.* As for nonautonomous beings – such as sentient nonhuman animals, fetuses, infants, and the severely retarded or severely demented – such discounting in accordance with the TRIA is, I suggest, rationally mandatory, in keeping with McMahan's view.

A Note about Medical Decision Making

In concluding this chapter, it is worth noting that we have examined the authority of advance directives from a strictly prudential standpoint. In thinking along these lines, and indeed in earlier outlining the two competing medical decision-making hierarchies, we have assumed the mainstream patient-centered perspective that dominates contemporary bioethics. Thus, we seek the patient's informed consent, consider the patient's advance directive, make a substituted judgment about what the patient would have wanted, or discern the patient's best interests. But there are substantial reasons to think that this patient-centered perspective and the associated decision-making hierarchies are morally oversimplified.[61]

Here I merely note the possibility that interests other than those of the patient may sometimes bear morally on the ethics of decision making for incompetent patients. For example, the decision to forgo life supports in accordance with an advance directive may be reinforced by consideration of family members' financial interests in limiting costly care as

---

[61] See, e.g., Thomas Mappes and Jane Zembaty, "Patient Choices, Family Interests, and Physician Obligations," *Kennedy Institute of Ethics Journal* 4 (1994): 27–46.

well as the hospital's or nursing home's interest in conserving human, technological, and financial resources. Suppose, on the other hand, that an apparently authoritative advance directive conflicts with such "external" moral considerations. Perhaps the former demands continued treatment, while other parties have an interest in discontinuing treatment – or vice versa. Although I would recommend staying close to the mainstream patient-centered ethic, I cannot defend this recommendation here.

6

Enhancement Technologies and Self-Creation

Alan recently turned forty. Although happy with his partner and content with his career, he ruminates about becoming middle-aged. His two daughters, ages eight and ten, talk in code around him, roll their eyes when he says something uncool, and seem to prefer their friends' company to his. Strangers call him "Mister." When Alan considers attending Arrowsmith and Fleetwood Mac concerts, he wonders how he would feel if he ran into teenage neighbors – "or would they not be caught dead there?" Inspecting himself in the mirror, he feels mediocre: double chin, protruding stomach, the V of his torso replaced with a U, plenty of gray infiltrating his brown hair. His days as a swimmer seem long gone. Alan decides to improve how he looks and feels by using Grecian Formula to restore his hair to its original color and starting an exercise regimen of running and weight training.

Having just finished college and taking time off before her first real job, Barbara is concerned about her physical appearance. She has dated little and is exasperated when male friends tell her that she is "a great conversationalist" and has a "terrific personality." Believing that her romantic prospects will brighten if she looks more alluring, Barbara joins a gym and decides to take a major step she has considered off and on since high school: breast enlargement with saline implants.

By nearly everyone's estimation, Carl is doing well as he approaches age thirty. He earns a good living as a real estate agent, has several close friendships, and greatly enjoys his pastimes of roller blading and genealogy research. These successes are especially impressive to those who know that he grew up in a highly dysfunctional family, in which he was subjected to neglect and emotional abuse yet was expected to take care of his siblings in a parental role. But even his friends' praise and support leave him unsatisfied. Carl often doubts his professional decision making and the way he presents living quarters to clients. His strong friendships notwithstanding, he sometimes bristles at perceived putdowns ("She didn't return my e-mail") or minor betrayals ("He didn't include me in the wedding party"). In his love life, he despairs at not finding a good partner, the women he dates consistently proving to be aloof, highly critical, sometimes a bit cruel. When he finds himself alone on a weekend, he feels empty and lonely. During these times certain persistent, even obsessional thoughts – about the possibility of tragedy befalling him or family members – nag at him (without really disrupting his functioning). After three sessions with a psychiatrist, who helps to clarify his psychological issues but concludes that he has no diagnosable illness or disorder, Carl decides that he wants changes in his life and, indeed, *in his personality*. He wants to be more confident, cheerful, and professionally decisive; less prone to feelings of being socially excluded, snubbed, or unworthy of a good partner; and less obsessive. Having read that selective serotonin reuptake inhibitors (SSRIs), such as Prozac and Paxil, often help with difficulties like his, Carl informs the psychiatrist he consulted that he would prefer a prescription for an SSRI to the course of psychotherapy the psychiatrist initially recommended. He has done plenty of psychotherapy already and believes an SSRI may provide the changes he seeks more quickly and less expensively than another round of therapy.

Delia is an ambitious, hard-working scholar of Mediterranean history. In midcareer, she is the author of two well-regarded books and looks forward to working on her third during an upcoming sabbatical. Since 2012, when the first genetic interventions for enhancement purposes received federal approval after years of clinical trials, Delia has toyed with the idea of undergoing two enhancements.[1] First, she would like to reduce her need for sleep. Today geneticists can introduce a gene into cells that can be implanted into a patient's hypothalamus. This gene causes the production of a protein that resets the suprachiasmatic nucleus (circadian clock), thereby reducing the need for sleep. Delia would also like to enhance her memory. It was recently discovered that, other factors being equal, an increase in the number of N-methyl-D-aspartate (NMDA) receptors in the brain improves long-term memory. Geneticists can now

---

[1] For a discussion that suggests that the two genetic interventions imagined here are realistic possibilities, see LeRoy Walters and Julie Gage Palmer, *The Ethics of Human Gene Therapy* (New York: Oxford University Press, 1997), pp. 102–6.

insert genes that encode these receptors into a patient's cells in order to increase the number of NMDA receptors in the brain. Delia is confident that these apparently quite safe genetic technologies will allow her to sleep several hours less per night and remember much more of what she reads, permitting her to work longer and more efficiently while also creating more waking time for leisure.

The four individuals just described seek to transform themselves in ways that, according to their own values, count as improvements. Alan wants to look and feel younger and more vigorous. Barbara wants to become more alluring. Carl wants to be more decisive and confident, less prone to worry and obsess. Delia wants to be able to remember more and sleep less. None of them is medically ill. Each of their enhancement projects involves the use of technologies, though Alan's Grecian Formula is so low-tech that we might forget that it is a technology. Interestingly, we are likely to regard his and Barbara's exercise as a credit to their own efforts and discipline more than to technology, even if they use cutting-edge weight-lifting equipment; accordingly, this component of their plans provokes little or no moral concern. On the whole, Alan's enhancement project seems the least troublesome morally. Barbara's interest in cosmetic surgery and Carl's intention to undergo "cosmetic psychopharmacology" are likely to give one greater pause. Meanwhile, Delia's project of genetic enhancement will strike many people as the most troubling of the four plans. But why are any of them troubling? Should they be? Before addressing these questions, let us examine the growth and character of enhancement technologies.

INTRODUCTION TO ENHANCEMENT TECHNOLOGIES

Let us use the term *enhancement technologies* to refer to *certain technologies when they are employed for purposes of enhancement*. Thus, the same technology – Prozac, for example – will count as an enhancement technology in certain contexts (when used for enhancement purposes) but not in others (when not use for enhancement). This slightly misleading phraseology is, I think, preferable to *biotechnological enhancements*, which is too specific for our purposes – excluding, for example, cosmetic surgery – and *technologies employed for enhancement purposes*, which is cumbersome. But unpacking enhancement technologies in terms of enhancements requires clarification of the latter.

In the biomedical context, enhancements are commonly understood as "interventions designed to improve human form or functioning beyond what is necessary to sustain or restore good health."[2] Put another way, enhancements are interventions to improve human form or function that *do not respond to genuine medical needs*, where the latter are defined (1) in terms of disease, impairment, illness, or the like, (2) as departures from normal (perhaps species-typical) functioning, or (3) by reference to prevailing medical understandings. This conception of enhancements identifies them by the goal of improvement in the absence of medical need. But sometimes enhancements are identified by the nature of their *means*. Some means of self-improvement, such as exercise, are considered natural and virtuous. By contrast, other means are perceived as artificial, as involving corrosive shortcuts, or as perverting medicine, rendering the self-improvement morally suspect and classifying it as an enhancement – for instance, steroid use to improve athletic performance. In this chapter, I will adopt the more common conception, which contrasts enhancements with *treatment* or *therapy*, interventions responding to genuine medical needs.[3] Accordingly, the intended uses of exercise and Grecian Formula, surgery, an SSRI, and genetic technologies in the vignettes all qualify as enhancements, for no medical illness or impairment motivates these interventions.

To be sure, one might question the meaningfulness of the treatment–enhancement distinction. Consider, for example, that children who are very short due to a deficiency in growth hormone (GH) are classified as receiving treatment in receiving synthetic GH. Meanwhile, children who have normal levels of GH, but are equally short simply because their parents are short, are classified as normal, so that their receiving synthetic GH counts as enhancement – despite their facing the same disadvantage in being short and standing to gain equally from the drug.[4] Also challenging the treatment–enhancement distinction is our vignette involving Carl, who is described as having no psychiatric illness or impairment. One might argue that, since he struggles with certain psychological

---

2 Eric Juengst, "What Does *Enhancement* Mean?" in Erik Parens (ed.), *Enhancing Human Traits* (Washington, DC: Georgetown University Press, 1998), p. 29

3 Regardless of how the concept of medical need is elaborated (see, e.g., Norman Daniels and James Sabin, "Clarifying the Concept of Medical Necessity," *Proceedings of the Group Health Institute* [Washington, DC: Group Health Association of America, 1991]: 693–707), there will be ambiguous cases in which it is difficult to determine whether a condition qualifies.

4 For background, see Gladys White, "Human Growth Hormone: The Dilemma of Expanded Use in Children," *Kennedy Institute of Ethics Journal* 3 (1993): 401–9; and, on the approval of the drug for normal short children, "FDA Approves Wider Use of Growth Hormone," *The Washington Post* (July 26, 2003): A12.

phenomena that can be ameliorated with medication, it means little to say that he is not ill, whereas someone who, say, barely qualifies as having depression or clinical anxiety is ill.[5]

In response to such skepticism, I acknowledge that the treatment–enhancement distinction will seem arbitrary in a certain range of cases because our distinction between normal and abnormal health will sometimes seem arbitrary. Nevertheless, I believe the distinction is compelling and meaningful in most contexts, the gray area between treatment and enhancement notwithstanding. For example, the cases of Alan, Barbara, and Delia are pretty straightforwardly cases of enhancement. (I think Carl's case also features enhancement, but I acknowledge its contestability.) Further, I am willing to accept the distinction for the sake of argument even if its status is questionable. Opponents of enhancements need the distinction in order to name the class of interventions they oppose. Because I will defend a cautious liberalism regarding enhancement technologies, I prefer to assume the conceptual field on which the debate over their appropriateness most naturally takes place – and then evaluate enhancements on moral and prudential grounds. So, hereafter, I will assume that *treatment* and *enhancement* are meaningful, contrasting terms.

While enhancement technologies are hardly new, the term *enhancement* is newer. Clinicians and ethicists began to express concerns about enhancements in the 1980s with the impending development of gene therapy: Should we employ genetic interventions, beyond therapeutic uses, to bring about other types of improvements such as changes in eye and hair color, increased height for those of ordinary stature, or greater intelligence for the cognitively normal?[6] Sensing the public's jitters about genetic technologies in general, many clinicians thought it wise to draw a clear line between genetic therapy, which the public was more likely to support, and genetic enhancement, which clinicians were willing to characterize as beyond the pale.[7]

---

[5] Had I wanted four *uncontroversial* cases of enhancement, I might have substituted for Carl someone who, despite concentrating quite well without medication, wants to take Ritalin simply in order to perform better on exams. But I think the ambiguity of Carl's situation is instructive. Also, Carl's self-conscious aim to change his personality with medication will no doubt persuade many clinicians that his case involves enhancement.

[6] Carl Elliott, *Better Than Well: American Medicine Meets the American Dream* (New York: Norton, 2003), pp. XVII–XVIII.

[7] See, e.g., French Anderson, "Human Gene Therapy: Scientific and Ethical Considerations," *Journal of Medicine and Philosophy* 10 (1985): 275–91. One source of public discomfort was the historical specter of various eugenics movements in the first half of the twentieth century: most notoriously, the Nazi campaign to eliminate Jews, Gypsies,

Today enhancements come in enormous varieties. We use school, educational videos, and academic summer camps to enhance intellectual capabilities. Exercise, a controlled diet, and tennis lessons serve to improve various forms of physical functioning. Fluoride in tap water enhances our teeth's ability to fight off tooth decay, and routinely administered immunizations and vaccines increase our immune system's ability to ward off illnesses. (Some, however, would not count fluoride and immunizations as enhancements, suggesting that they represent a distinct third category: *prevention*.[8]) In order to enhance our looks, we lift weights, dye our hair, regrow hair with Rogaine for Men, fight wrinkles with Botox, use makeup, and either avoid or seek sunlight in pursuing a desired complexion. Most of us drink coffee not only for enjoyment but also for greater alertness. Less everyday examples of enhancements include contact lenses that make green or brown eyes look blue, classes that reduce regional accents, and sex-change operations to make one more at peace with oneself (although some would argue that such surgery responds to a psychological disorder and therefore should count as treatment). While the range of enhancements is vast, this chapter will focus on three types of enhancement technologies: cosmetic surgery, cosmetic psychopharmacology, and genetic enhancements. A brief overview of the growth and character of these types of enhancements follows.

Of the three types, cosmetic surgery is most familiar. By *cosmetic surgery* I mean surgery whose purpose is primarily esthetic: to make someone more physically attractive. It includes breast augmentation, facelifts, liposuction, hair transplants for balding men, most cases of nasal reconstruction, and such extreme measures as reshaping the face to make it appear more Caucasian and removing a couple of ribs to make a woman look more like a supermodel. Cosmetic surgery does not include nasal reconstruction when the primary purpose is to correct a breathing problem; nor does it include breast reduction surgery for the sake of relieving strain on chest and back muscles. In wealthy countries, especially in North America, cosmetic surgery is big – and growing – business. Between 1992 and 1999, according to the American Society of Plastic Surgeons, the number of cosmetic surgeries performed in the United States and Canada rose

homosexuals, and the mentally "feeble," but also eugenics movements in the United States, Great Britain, and other countries. For a helpful discussion, see Allen Buchanan et al., *From Chance to Choice: Genetics and Justice* (Cambridge: Cambridge University Press, 2000), ch. 2.

[8] See, e.g., Erik Parens, "Is Better Always Good? The Enhancement Project," in *Enhancing Human Traits*, p. 5.

175 percent, with especially high increases in breast augmentation and liposuction.[9]

In his landmark book *Listening to Prozac*, psychiatrist Peter Kramer coined the term *cosmetic psychopharmacology* to refer to the use of psychotropic medications by individuals who are not really ill but want to become "better than well": more energetic, socially confident, and attractive in self-presentation.[10] Perhaps "better than adequate" more aptly captures how many of these individuals view their goal; though not medically ill, they often face too much frustration to feel especially well when they seek assistance. In any case, interestingly, these individuals sometimes want not only to feel better, but also to change their personalities in nontrivial ways.

Let's use the term *cosmetic psychopharmacology* a bit broadly to include any use of prescription or over-the-counter medications, herbs, and other substances whose purpose is to enhance mental and/or social functioning in individuals who have no psychiatric illness or impairment. Examples beyond the cases involving SSRIs are the use of Ritalin to improve cognitive performance (without treating a condition such as attention deficit/hyperactivity disorder [ADHD]) and the use of propranolol for reducing normal anxiety in order to enhance musical performance.[11] Admittedly, clinical creativity can produce a diagnosis for virtually anyone a psychotherapist would like to treat, a conceptual liberality driven partly by insurance companies' demand for a diagnosis as a condition for coverage or reimbursement. Thus Carl's psychiatrist might diagnose him as having an adjustment disorder, citing his impending thirtieth birthday as the relevant stressor.[12] But that should not prevent us from recognizing that some such individuals are not really ill even if they suffer from certain quirks and life challenges (assuming these do not seriously disrupt

9 Sara Goering, "The Ethics of Making the Body Beautiful: What Cosmetic Genetics Can Learn from Cosmetic Surgery," *Philosophy & Public Policy Quarterly* 21 (Winter 2001), p. 21. Goering cites www.plasticsurgery.org.

10 *Listening to Prozac* (New York: Viking, 1993). Seven years later, an entire issue of a bioethics journal was devoted to this topic: *The Hastings Center Report* 30 (2) (2000).

11 See, e.g., Lawrence Diller, "The Run on Ritalin: Attention Deficit Disorder and Stimulant Treatment in the 1990s," *Hastings Center Report* 26 (2) (1996): 12–18; Claudia Mills, "One Pill Makes You Smarter: An Ethical Appraisal of the Rise of Ritalin," *Report from the Institute for Philosophy and Public Policy* 18 (4) (1998): 13–17; Peter Whitehouse et al., "Enhancing Cognition in the Cognitively Intact," *Hastings Center Report* 27 (3) (1997): 14–22; and Jacquelyn Slomka, "Playing with Propranolol," *Hastings Center Report* 22 (4) (1992): 13–17.

12 American Psychiatric Association, *Desk Reference to the Diagnostic Criteria from DSM-IV* (Washington, DC: American Psychiatric Association, 1994), pp. 273–4.

functioning). Our discussion will focus on cases of cosmetic psychopharmacology, like Carl's, that involve deliberate changes of personality.

Like cosmetic surgery, cosmetic psychopharmacology appears to have
grown massively in recent years. I say "appears to" because it is much easier to find statistics on the overall use of certain medications like Prozac –
which we know, anecdotally from Kramer and others, are sometimes used
in the absence of a psychiatric illness – than to determine how often these
medications are used for such enhancement purposes. The greatly expanded use of SSRIs such as Prozac, Paxil, and Zoloft, and the increasing
number of officially recognized disorders for which they are prescribed,
strongly suggest an increase in cosmetic psychopharmacology. Today psychiatrists and primary care physicians prescribe antidepressants, including SSRIs, to treat not only depression but also social anxiety disorder,
panic disorder, paraphilias such as exhibitionism and fetishism, sexual im-
pulsivity, obsessive-compulsive disorder, and premenstrual dysphoric disorder.[13] The drugs' success in treating these disorders has been associated
with marked increases in estimates of how often these disorders occur.[14]
Another interesting case is that of stimulants such as Ritalin and Adderall.
These apparently safe medications help people who have ADHD as well
as people who simply have difficulty concentrating – although drawing
a meaningful, clear line between these two groups is hard at best. From
1990 to 1995, use of Ritalin in the United States increased by 500 percent. In 1995 2.6 million Americans were taking the drug, many of them
children. Interestingly, this half-decade saw no increase in Ritalin use in
any European or Asian country.[15]

By contrast with cosmetic surgery and cosmetic psychopharmacology,
genetic enhancement is not yet a technical, much less a clinical, reality. However, research that could lead to genetic enhancements is well
underway, and genetic therapy is common today. Some background on
genetic research and therapy will set the stage for thinking about possible
enhancements of the future.

[13] Elliot, *Better than Well,* p. 123.
[14] David Healy, "Good Science or Good Business?" *Hastings Center Report* 30 (2) (2000):
    19–22.
[15] See Elliott, *Better Than Well,* p. 249; and Lawrence Diller, *Running on Ritalin* (New
    York: Bantam, 1998), pp. 35–6. Diller, a pediatrician specializing in behavioral prob-
    lems, conjectures that fewer than half of the children for whom he prescribes stimu-
    lants meet clinical criteria for ADHD ("Prescription Stimulant Use in Children: Ethi-
    cal Issues," December 12, 2002, presentation to the President's Council on Bioethics
    [www.bioethics.gov], Washington, DC).

With a $3 billion budget, the (U.S.) National Institutes of Health and Department of Energy initiated the Human Genome Project in 1990. The project's goals were to identify the estimated 30,000–35,000 human genes and to sequence the roughly 3 billion base pairs of nucleotides (DNA subunits) so that scientists could proceed to translate that knowledge into clinical applications. The project was completed in April 2003.[16] Scientists continue to work on revealing the functions of particular genes and developing clinical applications. Many institutions are engaged in genetic research – some funded by the National Institutes of Health or government agencies in other countries, some privately funded with an eye toward entrepreneurial opportunities.

As genetic research advances, a fundamental ethical issue is what sort of research and subsequent clinical applications should be permitted. As mentioned, many clinicians publicly support genetic therapy but not enhancement. The latter, they argue, is much less pressing than the treatment or cure of medical illness (or the disposition to develop a specific illness).[17] Genetic enhancement is also likely to be technically more difficult, and therefore more dangerous, than therapy while posing the social dangers of valorizing some medically irrelevant traits over others (e.g., blonde hair over dark hair).[18] But increasingly many scholars have offered qualified endorsements for certain types of genetic enhancements.[19] Another frequently invoked distinction, which cuts across the therapy–enhancement divide, is that between somatic cell interventions and germline cell interventions. Genetic changes in somatic cells – such as skin, kidney, or muscle cells – will not be transmitted to offspring. By contrast, genetic changes to germline cells (gametes) – sperm or

---

[16] www.ornl.gov/TechResources/Human_Genome/project.
[17] Everyone agrees that genetic means of eliminating an individual's predisposition to schizophrenia or Huntington's disease would count as therapy. That is, *preventing* these diseases would be therapy as opposed to enhancement. By contrast, genetic means of strengthening one's immune system – which could, in turn, prevent certain illnesses – is more ambiguous between therapy and enhancement or even a separate category of prevention: One may prevent certain illnesses by improving one's immunity beyond (currently) ordinary human capacities.
[18] See, e.g., French Anderson, "Genetics and Human Malleability," *Hastings Center Report* 20 (1) (1990): 21–4.
[19] See, e.g., Walters and Palmer, *The Ethics of Human Gene Therapy*, ch. 4; Glenn McGee, *The Perfect Baby* (Lanham, MD: Rowman & Littlefield, 1997), ch. 7; Buchanan et al., *From Chance to Choice*, ch. 8; and Anita Silvers, "Meliorism at the Millennium: Positive Molecular Eugenics and the Promise of Progress without Excess," in Lisa Parker and Rachel Ankeny (eds.), *Mutating Concepts, Evolving Disciplines: Genetics, Medicine, and Society* (Dordrecht, the Netherlands: Kluwer, 2002): 215–34.

egg cells or their stem-cell precursors – will be transmitted to offspring. For this reason, germline genetic interventions are likely to have more far-reaching consequences, including any unforeseen deleterious effects, than somatic cell interventions. Not surprisingly, then, among the four categories of genetic intervention implied by the two intersecting distinctions, germline enhancement has the weakest public and academic support, while somatic cell therapy enjoys the broadest support – and is, to my knowledge, the only type of genetic intervention attempted so far.[20]

Before turning to types of genetic therapy and possibilities for genetic enhancement, let's briefly consider the mechanics of genetic interventions. How can genetic information be deliberately *put into* a human being? First, it is necessary to select target cells, to which genes will be directed. Next, a vector for directing this material is needed. For example, a gene could be inserted into a retrovirus (the vector), which is harvested and directed to a patient's white blood cells (the target) *ex vivo*, before the genetically "corrected" blood cells are reintroduced into the patient's body. More commonly today, genetic material is introduced directly into the body. Other vectors include different types of viruses as well as nonviral vectors such as artificial lipid spheres. These variations represent *gene addition*, currently the standard technique. But scientists hope to develop reliable techniques of either *gene correction*, in which a normal gene segment replaces the segment containing the defect, or *gene repair*, in which a normal piece of DNA is inserted into a cell and the cell's own repair mechanisms correct the erroneous DNA sequence.[21]

Somatic cell gene therapy is rapidly expanding.[22] The first human somatic cell gene therapy experiment began in 1990. The subjects were children with adenosine deaminase (ADA) deficiency disease, which harms the white blood cells – specifically, T cells – that fight infections, typically leading to death before age two years. In this experiment, the target was T cells and the vector was a retrovirus. Five years later, 100 protocols for gene therapy experiments had been federally approved in the United States or were under review by the Food and Drug Administration.

---

[20] Accidental transmission of genetic material to eggs or sperm has apparently occurred, however, due to imperfect control over where therapeutically introduced genes will go. See Nancy King, "Accident and Desire: Inadvertent Germline Effects in Clinical Research," *Hastings Center Report* 33 (2) (2003): 23–30.

[21] Mark Frankel, "Inheritable Genetic Modification and a Brave New World," *Hastings Center Report* 33 (2) (2003), p. 32.

[22] In this paragraph (except the last sentence) I have benefited from Walters and Palmer, *The Ethics of Human Gene Therapy*, ch. 2.

The target diseases of these protocols were various types of cancer (sixty-three protocols), human immunodeficiency virus (HIV) disease and acquired immune deficiency syndrome (AIDS) (twelve), genetic diseases (twenty-two) including cystic fibrosis and Gaucher disease, and miscellaneous others (three). As of July 2004, according to one highly respected database, there were 987 approved, ongoing, or completed gene therapy trials worldwide.[23]

If genetic enhancements are ever permitted, or attempted without being permitted, they may include those mentioned in the vignette involving Delia, the history scholar: interventions aimed at reducing the need to sleep and improving long-term memory. Another possibility is enhancement of oxygenation mechanisms beyond that needed for good health. Geneticists might also improve the power and resilience of muscles above the normal level by applying to unafflicted individuals the same genetic techniques used to promote muscle growth in patients with muscular dystrophy. Many athletes would welcome such advantages as the last two. A cosmetic example of genetic enhancement is a cure for baldness that counteracts the gene(s) for male-pattern baldness with a cream that carries hair-growth genes in artificial lipid spheres.[24] Regarding a gene that, when mutated, causes a disease characterized by enormous appetite and overeating, scientist Hunt Willard offered this speculation: "If one copy of that [normal] gene keeps people from overeating, maybe two or three copies . . . will be the new Jenny Craig" (referring to a popular weight loss plan).[25] More disturbingly, since genes for skin pigmentation have been identified, one can imagine requests for genetically changing this aspect of an individual's racial appearance. As a final example, some scientists believe that genetic means of postponing normal aging, thereby lengthening a person's life span, may not be far off.[26] It need hardly be said that many people would be interested in that technology.

---

[23] www.wiley.co.uk/genmed/clinical (provided by *The Journal of Gene Medicine*).

[24] Rick Weiss, "Gene Enhancements' Thorny Ethical Traits," *The Washington Post* (October 12, 1997), p. A19.

[25] Ibid., p. A18.

[26] For insightful discussions of the ethical issues provoked by this possible technology, see Eric Juengst et al., "Biogerontology, 'Anti-Aging Medicine,' and the Challenges of Human Enhancement," *Hastings Center Report* 33 (4) (2003): 21–30; and David Gems, "Is More Life Always Better? The New Biology of Aging and the Meaning of Life," *Hastings Center Report* 33 (4) (2003): 31–9. Remarkably, as Gems explains, the biological processes associated with aging may not be inevitable, opening up the possibility of treating these processes as a disease or cluster of diseases – in which case antiaging efforts would qualify as treatment, or perhaps prevention, but not enhancement.

Are genetic enhancements likely to emerge? One might think that widespread fear and disapproval of this technical frontier would translate into effective legislative barriers. I suspect, however, that such negative reactions will be less common as the public reflects on some of the more benign possibilities. For example, if we count improving the immune system as an enhancement, then such improvement by genetic means is a promising candidate for an enhancement that would eventually earn widespread public approval. Among those who are open to genetic enhancements, many ethicists and clinicians are likely to prioritize health-related enhancements – such as improving an already normal ability to fight infections (assuming this counts as an enhancement) or improving a person's already normal oxygenation mechanisms – over those serving primarily cosmetic purposes.[27] In any event, market pressures may overwhelm widespread disapproval if it persists. In the words of French Anderson, a preeminent geneticist who has long opposed the development of genetic enhancements, "[it] is going to happen. Congress is not going to pass a law keeping you from curing baldness."[28]

Perhaps the main route to the emergence of genetic enhancements will parallel those leading to cosmetic surgery and cosmetic psychopharmacology: the expanding use of interventions approved and developed for therapeutic purposes.[29] Thus Delia's reduced-sleep genetic enhancement may be a by-product of research aiming at therapies for circadian rhythm disorders, while the intervention used for memory enhancement may originally be approved as genetic therapy for Alzheimer's disease. Medical rationales often follow the new uses, which are sometimes defended as preventive measures. In this way, today drugs to treat Alzheimer's disease are used earlier and earlier either to "treat" what is called *mild cognitive impairment* or to "prevent" the onset of the disease by delaying the (hitherto normal) process of brain aging. Naturally, the conceptual difficulties of cleanly distinguishing treatment and enhancements facilitate such trends.

The remainder of this chapter will examine enhancement technologies with special attention to human identity. First, it will address relatively familiar concerns about enhancement technologies. Then, after exploring the relationship between enhancement technologies and identity, the

---

[27] See, e.g., Walters and Palmer, *The Ethics of Human Gene Therapy*, ch. 4; and Silvers, "Meliorism at the Millennium."

[28] Quoted in Weiss, "Gene Enhancements' Thorny Ethical Traits," p. A18.

[29] This paragraph is largely due to Eric Juengst.

chapter will present and address the identity-related charges (1) that certain enhancement projects are *inauthentic* and (2) that certain enhancements *violate inviolable core characteristics*. The discussion will conclude on a note of cautious openness about the use of enhancement technologies in self-creation.

FAMILIAR CONCERNS ABOUT ENHANCEMENT TECHNOLOGIES[30]

Several prudential, moral, and social concerns about enhancement technologies are relatively familiar. Some focus on the morality of individual choices, such as those of prospective users of enhancement technologies or their physicians. Others focus on society's obligations and/or the most defensible public policies. Our somewhat compressed discussion of these well-known concerns will (tentatively) recommend cautious openness to enhancement technologies – with one important justice-based caveat. Except where otherwise indicated, we will consider enhancements for competent adults, for whom the concept of self-creation is most relevant.

(1) Complicity with Morally Problematic Social Norms. This issue concerns the morality of individual choice. Cosmetic surgery, cosmetic psychopharmacology, and genetic enhancements promote some very troubling social values. Breast augmentation promotes sexist attitudes about how women should look: as sexually pleasing to men as possible, even if looking this way imposes significant risks and costs. Meanwhile, part of what drives Carl's enhancement project is his desire to be more effective at work and his longing for a more attractive personality. (Another chief, and morally unproblematic, motive is his desire to feel better.) Since Carl is already professionally successful and has strong friendships, one might think his desire for enhancement reflects our culture's tendency to valorize hypercompetitiveness and "designer" personalities. In a more obvious way, Delia's enhancement project also promotes hypercompetitiveness (although she is interested in having more free time as well). So at least three of the enhancement projects raise the issue of complicity with morally problematic social norms.[31] Is it ethically problematic for the

---

[30] Several arguments in this section were sketched in my "Prozac, Enhancement, and Self-Creation," *Hastings Center Report* 30 (2) (2000), pp. 38–9.

[31] This theme has been extensively explored in connection with cosmetic surgery. See, e.g., Margaret Olivia Little, "Cosmetic Surgery, Suspect Norms, and the Ethics of Complicity," in Parens, *Enhancing Human Traits*: 162–76; Susan Bordo, "*Braveheart, Babe,* and the

three patients to seek these enhancements and wrong for clinicians to provide them?

In view of important differences among enhancement technologies and the circumstances of those seeking to use them, I suggest that the only plausible way to address this question is case by case. For every case, two questions are especially salient. First, *to what degree is use of a particular enhancement technology tied to and supportive of a clearly unjustified social norm?* This, in turn, requires us to ask both how clear it is that the norm is unjustified, and how the use of the technology in question is socially related to that norm. Second, *to what degree, if at all, would an individual experience suffering, lost opportunity, or other harm as a result of not using the technology?*

Regarding the first question, the relationship between breast enlargement for cosmetic purposes and sexist expectations for women's bodies is rather clear.[32] By contrast, the relationship between unjustified social norms and Carl's and Delia's plans is less clear. For among their various reasons for seeking change are some that are clearly beyond reproach. Moreover, it is far less obvious that the culprit norms – promoting hypercompetitiveness in both enhancement projects as well as a decisive, confident personality in Carl's – are morally problematic.

The second question, about the personal cost of foregoing an enhancement, requires us to look closely at an individual's circumstances and the possibility of acceptable alternatives that reduce or eliminate the cost. For example, someone who desires a major change in physical appearance may be more satisfied in the long run by eliminating, through psychotherapy, the insecurity underlying the desire. Perhaps psychotherapy is more effective than Rogaine in achieving long-term satisfaction for balding men. Then again, sometimes the personal costs of bucking social pressures to embrace some enhancement may be high, especially where there are no viable alternatives. A possible example is orthodontics for very unsightly but functional teeth.

Contemporary Body," in Parens, *Enhancing Human Traits*: 189–221; and Goering, "The Ethics of Making the Body Beautiful."

[32] However natural the preference for large breasts may seem, it is apparently not universal in view of the preferences expressed through Renaissance art. Moreover, as feminist scholarship suggests, history reveals an insidious tendency of men to regard women as valuable primarily as mothers, caregivers, and, more saliently here, men-pleasers: "the norms of a good woman, unlike those of a good man, tend to value her function for others: [she] is nurturing of others and beautiful for them to behold" (Margaret Olivia Little, "Why a Feminist Approach to Bioethics?" *Kennedy Institute of Ethics Journal* 6 [1996], p. 6).

Because the two questions should be addressed case by case, legitimate concerns about complicity are compatible with some people's use of enhancement technologies and some clinicians' making them available. At the same time, it is incumbent on clinicians not to diminish patients' autonomy by preying on their insecurities, distorting facts, or failing to discuss major alternatives to their services.

(2) Promoting Biopsychiatry's Reductionist Agenda. Addressing cosmetic psychopharmacology in particular, some critics charge that use of SSRIs and similar drugs (whether for treatment or enhancement) promotes biopsychiatry's dubious agenda of reducing emotional and personal struggles to mechanistic terms – as if these struggles were just another form of pain to be treated with pills. According to these critics, this agenda threatens our self-conceptions as reasonable agents.[33]

Nevertheless, people ought to be free to agree with this agenda. Such reductionism is one of many contested philosophical positions with regard to the nature and extent of our freedom. Further, even if we agree on the practical importance of understanding ourselves as responsible agents, cosmetic psychotherapy does not preclude an appropriate relationship to this value. At some level, surely, we are responsible agents. But we are also feeling creatures. Suspiciousness, self-esteem problems, and obsessiveness are connected with our agency, but they are also closely tied to unpleasant experiences, which medications may help to alleviate. Besides, Carl's self-creation project is a powerful expression of his own agency even if a drug plays a central role in this project.

(3) Fostering Social Quietism. Also applying to cosmetic psychopharmacology in particular is the concern that it encourages social quietism: Patients may accept drug-induced complacency over active struggle to change the social conditions that contribute to their discontent, leaving these problems untouched.[34] But, while there may be some risk of social quietism, the risk accompanies all uses of mood-improving drugs – not just cosmetic psychopharmacology – as well as mainstream religions and many other clearly acceptable practices and institutions that may brighten our outlooks. Additionally, if cosmetic psychopharmacology improves someone's

---

[33] See, e.g., Carol Freedman, "Aspirin for the Mind? Some Ethical Worries about Psychopharmacology," in Parens, *Enhancing Human Traits*: 135–50.

[34] See Carl Elliott, "The Tyranny of Happiness: Ethics and Cosmetic Psychopharmacology," in Parens, *Enhancing Human Traits*, p. 180; and Diller, "The Run on Ritalin," pp. 14–15.

spirits, that may free him to act on his deepest convictions rather than assuming that such effort is hopeless.

(4) Self-Defeating and Potentially Coercive Enhancements. In competitive contexts, problems arise if nearly everyone avails herself of the relevant enhancement. First, the result is self-defeating, because each user of cosmetic psychopharmacology or genetically enhanced memory, say, will bear the expense and other personal costs of the intervention without gaining a competitive advantage.[35] To cite a familiar nontechnological example of this phenomenon, most law school applicants take an expensive, time-consuming Law Scholastic Aptitude Test (LSAT) prep course simply to keep pace with each other; they do not gain a competitive advantage. And if most people grow taller due to genetic enhancement, they do not thereby gain any edge. A second problem is that those who would prefer not to use a particular technology may feel social pressure, perhaps even coercion, to do so for fear of falling behind competitors.[36]

These concerns about self-defeating uses of enhancement technologies and overwhelming social pressures to conform are speculative with regard to cosmetic psychopharmacology and genetic enhancement. By contrast, it may already be true (though I am not aware of any rigorous evidence) that women in some professional contexts, such as popular music and the film industry, have begun to feel significant social pressures to have breast enlargement surgery. The solution to this potentially serious problem is not obvious, just as the solution to the LSAT prep course problem remains unclear. It is noteworthy, though, that some of the goods sought in competitive contexts have intrinsic, and not only competitive, value to individuals – reducing concerns about self-defeating widespread use. For example, even if (quite unrealistically) every history scholar genetically enhanced her ability to function well on less sleep, so that none gained a competitive advantage, all would benefit by having more discretionary time. On the whole, the present concerns do not seem to cast significant moral doubt on enhancement projects such as those featured in our vignettes. They may, however, cast significant doubt on certain enhancements in particular contexts, such as steroid use in sports, as discussed later.

---

[35] Dan Brock, "Enhancements of Human Function: Some Distinctions for Policymakers," in Parens, *Enhancing Human Traits*, p. 60.

[36] See Diller, "The Run on Ritalin," p. 16; and Slomka, "Playing with Propranolol," p. 15.

(5) Safety. Prominent in discussions of enhancement technologies are the risks to prospective users. But what risks an intervention poses is an empirical question. Because some of the interventions in question are either relatively new, like SSRIs, or still unrealized, as with genetic enhancements, the extent of the risks may be unknown. Meanwhile, even known risks may be hard to discern amid the glitter of celebrated benefits. This point highlights the paramount importance of an informed consent process that includes a thorough and balanced – that is, autonomy-promoting – discussion of risks. It does not justify paternalistic prohibition of the use of these technologies, except in rather extreme cases featuring great risk and little prospect of compensating benefit. As with several other legitimate concerns, we should address concerns about safety case by case.

"But," one might reply, "this begs the question by assuming that safety concerns do not justify a blanket prohibition of certain categories of enhancements or, more radically, all medical enhancements." But if I succeed in showing that there is insufficient reason to rule out certain categories of enhancements across the board, this will undermine a fortiori the more radical claim that all medical enhancements should be prohibited on the basis of their risks. Here I will focus on genetic enhancements, the arguments for prohibiting which seem stronger and more widely accepted than the arguments for prohibiting cosmetic surgery or cosmetic psychopharmacology.

Those who oppose genetic enhancements on grounds of safety emphasize the absence of a precise vector for introducing genetic material into patients' bodies. The use of imprecise vectors, which deposits desirable genes while leaving undesirable genes in place, creates much uncertainty about what will happen in a patient's body. This opens the door to unanticipated genetic repercussions with unknown risks. Concerns about unknown risks, which are sobering in the case of somatic-cell interventions, are especially grave in the case of germline interventions because any deleterious genetic effects will be passed on indefinitely to future generations. But, in both types of attempted enhancements, somatic-cell and germline, the risks are likely to outweigh the benefits, according to the argument. That is because the cost of not pursuing the enhancement – accepting normal but suboptimal functioning – is not especially great. By contrast, the cost of not pursuing genetic *treatment* – accepting serious disease or impairment – is sometimes very great; absent alternative treatments, the prospect for benefit often yields an acceptable risk–benefit ratio. Thus, the argument concludes, sound risk–benefit assessment justifies a prohibition of genetic enhancement but not genetic treatment.[37]

This argument against genetic enhancements is oversimplified. Even if *on average* the most promising but untried genetic enhancements will offer less favorable risk–benefit ratios than the most promising but untried genetic treatments, it hardly follows that no prospective genetic enhancements will offer an acceptable risk–benefit ratio. And since we are considering competent adults making self-regarding decisions, a liberty we tend to prize very highly in biomedicine, a blanket prohibition is less compelling than a willingness to examine proposals for enhancement case by case. It is worth noting that the emphasis on crude vectors concerns the status quo of gene addition. But geneticists might develop more precise methods of gene correction and gene repair.[38] Only with a much greater understanding of the realistic possibilities can we assess the possible benefits of genetic enhancement. And only with such information can we make intelligent risk–benefit assessments. Since the question is empirical, there is a strong case for permitting cautious research on techniques that might lead to genetic enhancements.

Currently we know much more about the risks and benefits of the other two types of enhancement technologies on which we have focused. The benefits of cosmetic surgery in terms of improved self-esteem are as familiar as the risks. Meanwhile, cosmetic psychopharmacology's benefits are becoming familiar; its risks are the same as those for psychotropic drugs used in treating mental illness, which have been studied extensively. The case for making cosmetic surgery illegal or strongly discouraging it across the board with professional guidelines *in the name of safety* seems fairly weak. This leaves open the possibility, however, that a particular type of cosmetic surgery (say, rib removal) offers such a poor risk–benefit ratio that professional bodies should discourage or prohibit its members from offering these services; in truly extreme cases – say, elective limb amputation for someone who wants to be an amputee[39] – perhaps these surgeries should be outlawed. Similarly, cosmetic psychopharmacology

---

[37] See, e.g., Glannon, *Genes and Future People*, ch. 3; and Anderson, "Genetics and Human Malleability."

[38] See, e.g., Andy Coghlan, "'Editing' Fixes Disease Gene," *New Scientist* (July 12, 2003): 3.

[39] Elliott explores this extraordinary phenomenon in *Better Than Well*, ch. 9. A question that immediately arises about such amputation "wannabees" is whether they suffer from a treatable psychiatric illness. It is difficult to imagine a healthy person autonomously choosing to lose an arm or leg when there is no medical condition calling for amputation.

does not deserve to be professionally discouraged or outlawed *for reasons of safety*, except possibly in cases featuring exceptionally poor risk–benefit ratios.

(6) Autonomy. Even if an intervention poses a reasonable risk–benefit ratio, there remains the important moral question of whether a person's consent to the intervention is autonomous. Did she sufficiently understand the nature of the procedure, its foreseeable risks and benefits, and alternative means to achieve similar benefits? Was her choice sufficiently voluntary? Would full awareness of the social forces influencing her decision lead her to decide differently? (See Chapter 3's analysis of autonomy.)

At first glance, it would seem that such questions are answerable only case by case. Some critics of enhancement technologies disagreee, however. In their view, the autonomy of prospective enhancement users – or users of certain kinds of enhancements such as cosmetic surgery – is systematically undermined by morally obnoxious social norms (e.g., that a small-breasted woman is inadequate) and aggressive advertising that promotes such norms while misleadingly portraying the enhancement. Combining these concerns, Naomi Wolf writes:

When a modern woman is blessed with a body that can move, run, dance, play, and bring her to orgasm; with breasts free of cancer, a healthy uterus, a life twice as long as that of the average Victorian woman, long enough to let her express her character on her face; with enough to eat and a metabolism that protects her by laying down flesh where and when she needs it . . . the Age of Surgery undoes her immense good fortune. It breaks down into defective components the gift of her sentient, vital body and the individuality of her face, teaching her to experience her lifelong blessing as a lifelong curse.[40]

There is much wisdom in this assessment. In a culture in which women practically have to be models in order to succeed in Hollywood or the popular music industry, a culture in which magazines from *Playboy* to *People* and television shows from *Baywatch* to *Friends* continually up the ante for what counts as an acceptable female physique, cosmetic businesses are well positioned not only to respond to consumers' preferences but also to shape those preferences.[41] Provoking insecurities about one's

---

[40] *The Beauty Myth* (New York: Perennial, 1991), p. 99. Quoted in Goering, "The Ethics of Making the Body Beautiful," p. 23.

[41] Focusing on the movie and advertising industries, Bordo also argues that women's choices about using enhancements are seriously constrained ("*Braveheart, Babe*, and the Contemporary Body"). For an argument that the option of cosmetic surgery empowers

looks is big business. Indeed, Miller, Brody, and Chung have argued, with extensive documentation, that cosmetic surgeons' marketing routinely violates the Code of Ethics for the American Society of Plastic and Reconstructive Surgeons; for instance, their advertisements often create false or unjustified expectations of benefits and appeal primarily to laypersons' anxieties or emotional vulnerabilities.[42]

But, in my view, these concerns about the autonomy of individual choices about cosmetic surgery – concerns that might sometimes extend to other sorts of enhancements – do not undermine the proposal to evaluate critically the uses of enhancement technologies case by case. Assuming Chapter 3's analysis of autonomy is on the right track, we should judge that cultural norms for large breasts, while posing challenges to autonomous choice, do not preclude it. On the informational side, we should demand that businesses market their products responsibly, as many cosmetic surgeons have apparently not done. Methods and the contents of advertising can be more or less autonomy-promoting. As for voluntariness, some decisions to use enhancement technologies will be genuinely voluntary and some will not, depending on the agent's psychological profile and circumstances. In sum, the concern about autonomy motivates a critical examination of the messages and methods of advertising, as well as the messages that people and institutions convey throughout our culture. This concern does not persuasively argue against the use of enhancement technologies in general, against a broad category of enhancement such as cosmetic surgery, or against the case-by-case assessment of these technologies.

(7) Transgressing the Bounds of Medicine. Another objection to enhancement technologies is that providing them oversteps the proper bounds of medicine. According to this view, medicine involves all and only efforts to treat disease and impairment, so medical practitioners should confine their professional activities to these efforts and not offer enhancements. While reconstructive surgery falls within the proper bounds of medicine, cosmetic surgery does not. While psychopharmacology aimed at the treatment of mental illness or impairment is appropriate, cosmetic psychopharmacology is not. And while genetic treatment falls within the

women more than it constrains them, see Kathy Davis, "The Rhetoric of Cosmetic Surgery: Luxury or Welfare?" in Parens, *Enhancing Human Traits*: 124–34.

[42] "Cosmetic Surgery and the Internal Morality of Medicine," *Cambridge Quarterly of Healthcare Ethics* 9 (July 2000), pp. 360–2; American Society of Plastic and Reconstructive Surgeons, *Code of Ethics* (revised 1992).

socially mandated job description of the medical profession, genetic enhancement does not.

According to one reply to this criticism, disease and impairment are value-laden social constructions rather than objective phenomena, casting doubt on their authority in demarcating the boundaries of medical practice. Thus masturbation and homosexuality used to count as medical problems but no longer do; anticommunism was treated as a mental illness in Soviet hospitals but no longer is.[43] This reply strikes me as unpromising, however. Neither masturbation, nor homosexuality, nor anticommunism was ever an illness – and their being treated as such reflected the unenlightened or disingenuous practices of their medical eras.[44] Disease – like its opposite, health – is a biological reality, interfering with an organism's capacity to survive and function. But to acknowledge the reality of health and disease is not to deny the existence of a gray area between them in which line-drawing feels arbitrary. Oppositional concepts can be reality-based without being neatly delineated. Anyway, even if I am wrong that disease is a biological reality, the crucial point is that my position does not depend on the thesis that disease is a social construction.

A more promising reply to the charge of overstepping the bounds of medicine challenges the premise that medical practitioners should offer only those services that respond to disease or impairment. This more promising approach need not insist that insurance programs should reimburse practitioners for providing enhancement services; it need only claim that for practitioners to provide such services does not ipso facto violate their professional obligations. Consider several examples.

Immunizations clearly lie within the bounds of acceptable medical practice, yet they enhance the immune system's ability to fight certain diseases rather than treating a disease or impairment. Once again, an opponent of enhancements might invoke a tripartite division of treatment, prevention, and enhancements, claiming that the proper bounds of medicine encompass only treatment and prevention and that immunization counts as prevention. Those sympathetic to this move should consider several other examples. Most of us believe that physicians' acts of abortion, sterilization, or prescribing contraceptives – or at least one or two of these – are legitimate medical services. Yet these services do not fall within the bounds of medicine as defined by the criticism, because

---

[43] See, e.g., Tristam Engelhardt, "The Disease of Masturbation: Values and the Concept of Disease," *Bulletin of the History of Medicine* 48 (1974): 234–48.

[44] Cf. Buchanan et al., *From Chance to Choice*, p. 122.

reproductive capacities are normal functions; contraceptives, sterilization, and abortion interfere with normal functioning without treating
an illness or impairment. This does not prevent most of us from thinking that doctors appropriately provide these services. Another example
is psychotherapy for someone who has no mental illness or impairment,
but is simply very troubled by certain life issues. While it may be reasonable to claim that insurance schemes should not have to cover such
psychotherapy, it is unreasonable to insist that psychiatrists and other
therapists should abstain from providing it as a professional service. The
present criticism of enhancements draws the bounds of appropriate medical practice too narrowly.

(8) Special Concerns about Children. So far, our discussion has focused on
the possibility of adults electing to use enhancement technologies and
professionals providing such services. Special concerns arise in the case
of children due to the fact that others, their parents or guardians, must
make decisions for them.[45] Whereas competent adults, as autonomous
choosers, may permissibly accept certain risks for themselves, the risks
they may take on behalf of children within their charge are much more
significantly circumscribed by the parental duty to foster the best interests
of their dependents.

Notably, this duty will require special attention to (1) children's safety
and (2) their right to an open future.[46] Regarding safety, risks will be
acceptable only if the prospects for benefits entail a better risk–benefit
ratio than that offered by available alternatives, including the alternative
of doing nothing. This point alone will make it very difficult to justify children's participation in research on risky enhancements, yet such research
would be a moral prerequisite to any clinical use of the enhancement (although, as noted earlier, sometimes enhancement technologies reach
the market despite having been studied only in contexts in which they
functioned as therapies). In general, therapeutic research – research in
which there is a nontrivial prospect of direct medical benefit to subjects
themselves – is easier to justify. Frequently there is no effective treatment

---

[45] For a sensitive discussion, see ibid., ch. 5.
[46] Regarding this right, see Joel Feinberg, "The Child's Right to an Open Future," in William
Aiken and Hugh LaFollette (eds.), *Whose Child?* (Totowa, NJ: Rowman & Littlefield,
1980). Two further issues – moral education versus the medicalization of misbehavior
and the meaning of children's performance – are thoughtfully discussed in the President's Council on Bioethics, *Beyond Therapy: Biotechnology and the Pursuit of Happiness*
(Washington, DC: PCB, October 2003), pp. 91–4.

for some condition afflicting children, and the status quo of nontreatment is unacceptable. But in cases of enhancement there is no disease or impairment whose correction is imperative; the status quo of a suboptimal form or functioning is not so bad as to justify risky research. At the same time, it might be premature to assume that *no* enhancements research on children is compatible with the best interests standard. Perhaps some prospective enhancements are exceptionally promising and pose little or no risk.

Unfortunately, some enhancements for children will come into use without any research on their use *as enhancements*. Rather, an intervention studied in contexts where it is therapeutic will eventually be used "off label" for enhancement purposes. This slippage is dangerous insofar as the benefits of enhancement will often be lower than those of therapy, with no proportionate decrease in risks.[47]

The right to an open future establishes a different requirement for parental decision making. Imagine that privately funded, renegade research on young adults has established the safety of a genetic intervention that destroys someone's ability to read. (It is safe in the sense that it achieves its aim without producing harmful side effects.) While most people would consider the destruction of one's ability to read a terrible harm, this group considers it an enhancement because they regard reading as fostering intellectual activity threatening to their faith tenets. A couple now requests this procedure for their fifteen-year-old son. Even if the teenager assents as autonomously as possible for someone of his age, parental consent for this procedure would seem wrong. The most salient reason is that losing the ability to read would drastically reduce this young person's range of opportunities in the future. That is, it would violate his right to a reasonably open future. Any acceptable enhancements for children must respect this right while meeting appropriate requirements of safety.

(9) Justice. Enhancement technologies are not available to all who might want them, raising concerns about justice. If Delia's insurance program won't cover genetic enhancements that reduce her need to sleep and improve her memory, as a tenured college professor she may be able to afford them. Commanding less discretionary cash are many unemployed or underemployed history Ph.D.s who compete with the likes of her for publications in prestigious journals: no genetic enhancements for them.

---

[47] See, e.g., White, "Human Growth Hormone."

Meanwhile, Carl can afford an SSRI if his insurance company will not cover it, but many less wealthy individuals cannot. Naturally, the same is true of Barbara's cosmetic surgeries, reminding us that this concern about justice has been with us for a long time.

Elsewhere I have argued what I will assume here: that every relatively wealthy nation has a justice-based obligation to ensure its citizens universal access to health care.[48] Right now the United States is nowhere near satisfying this obligation. Presumably, more of the same – that is, further dividing the haves and have-nots with new enhancement technologies available only to the advantaged – would increase the injustice. Does that mean that the United States should prohibit new markets, and perhaps even work to eliminate existing markets, in biomedical enhancements?

Not necessarily. It might be inconsistent to prohibit *on grounds of justice* the purchase of certain enhancements, such as those considered here, while permitting the purchase of other enhancements – especially those, such as first-rate private school educations, that *greatly* advantage the already advantaged in comparison with the least advantaged. On the other hand, perhaps we should counter injustice wherever we can, even if the political status quo prevents us from addressing it in all arenas. Being somewhat inclined to think so, I regard this justice-based concern as a significant challenge to biomedical enhancements, especially to their further development. Note, however, that this challenge concerns a society's policies, not the question of whether individuals can appropriately purchase or otherwise procure an available enhancement whose distribution is unjust.[49]

In response to this challenge, I offer several suggestions. First, universal access to (adequate) health care should be the foremost biomedical priority in the United States and any other wealthy nation that currently lacks universal access. This is more important than developing esoteric

---

[48] "Grounding a Right to Health Care in Self-Respect and Self-Esteem," *Public Affairs Quarterly* 5 (1991): 301–18 and "Why the United States Should Adopt a Single-Payer System of Health Care Finance," *Kennedy Institute of Ethics Journal* 5 (1996): 145–60.

[49] I suggest that one can sometimes responsibly make use of an enhancement – for oneself or one's dependent – even though its current distribution is unjust. For example, one can responsibly attend, or have one's child attend, an elite private school despite the distributive injustice that plagues our educational system. While such a choice makes one somewhat complicit in injustice, we enjoy rather broad liberty rights to pursue what is best for ourselves, and parents have an obligation to pursue what is best for their children (within reasonable limits). But how best to resolve the moral tension between avoiding complicity and pursuing what's best for oneself or one's child is not obvious.

enhancements or preserving markets for existing ones. Indeed, for any such country, I would recommend – although I cannot adequately defend this recommendation here – a moratorium on the development of new biomedical enhancements that confer significant competitive advantages (e.g., certain genetic enhancements) until universal access to health care has been achieved. I also favor, in the meantime, whatever can be reasonably done to discourage the injustice-exacerbating proliferation of currently available biomedical enhancements that confer major competitive advantages. For example, professional bodies might strongly discourage and perhaps even prohibit (whether or not they can effectively enforce) the prescription of Ritalin for children who do not have ADHD or another genuine disorder.

Second, justice in health care access is compatible with the availability of certain biomedical enhancements to those able and willing to pay for them. When everyone has access to needed health care, disparities resulting from a residual free market will be tolerable, I think, from the standpoint of justice. Importantly, though, the content of obligatory universal health care includes access to services that are reasonably effective in treating diseases or impairments (assuming resource constraints within a just system permit treating the condition in question[50]), but probably not access to clear-cut cases of enhancements. (We have noted the difficulty of deciding whether, for example, interventions to improve the body's ability to fight infections should count as enhancements. They are not clear-cut cases of enhancement but would likely fall within a health care entitlement. Thus we might revise our statement to read "treating *or preventing* diseases or impairments.") Whether an obligatory health care package should include coverage for sterilization, abortion, and certain other services that do not respond to medical illness, yet do not seem to be enhancements either, is more debatable.

Finally, society should vigorously work to eliminate injustices in other areas. For example, although the U.S. public school system affords universal access to education, it is far from providing universal access to *adequate* education. Consequently, the great disparities in quality of the schools educating American children result in highly inequitable impacts on lifetime opportunities. Raising inadequate public schools to adequacy is an urgent goal; and achieving this goal would arguably render the added benefits of the best private education tolerable from the standpoint of justice.

---

[50] This clause is close to one articulated in Buchanan et al., *From Chance to Choice*, p. 121.

In conclusion, familiar concerns about biomedical enhancement technologies motivate a critical-minded, case-by-case assessment of their provision and use. The concern about justice arguably calls for universal access to health care as a moral prerequisite – already satisfied in most industrial countries – to the development of new biomedical enhancement technologies that confer significant competitive advantages. But, contrary to some critics, these concerns do not justify permanently prohibiting the use of such technologies or large categories of them such as genetic enhancements.

Meanwhile, some concerns may justify prohibiting certain technologies in particular contexts. For example, the use of steroids in professional sports and the Olympics is prohibited – appropriately, I think – due to the very serious risks steroids pose as well as the potentially coercive pressure many athletes would feel to take these risks if steroid use were officially permitted. Moreover, participation in a professional sport or the Olympics implies acceptance of rules, such as the prohibition of steroid use, whose basis is a widely shared sense of fair competition and the spirit of particular athletic activities. To break one of these rules is to cheat twice: formally violating the rules and materially violating a public, shared understanding of the spirit of the competition.[51]

The possibility of prohibiting or strongly discouraging particular enhancements in specific contexts is consistent with the position tentatively recommended here: openness toward enhancement technologies, case-by-case assessment, and possibly universal access to health care as a prerequisite for further development of enhancement technologies that confer significant competitive advantages. Nevertheless, one could accept every argument presented so far in this chapter while maintaining that enhancement technologies, or at least some broad categories of them, are *inherently* problematic for reasons connected with personal identity. Before turning to this set of concerns, let us do something that critics advancing them have generally not done: clarify the relevant sense of identity.

THE RELATIONSHIP OF ENHANCEMENT PROJECTS TO IDENTITY

Enhancement projects, such as those characterized in our vignettes, seem to be connected with human identity – with who someone is, or, relatedly, who she wants to become. Carl Elliott even suggests that the most

---

[51] For a thoughtful set of reflections, see President's Council on Bioethics, *Beyond Therapy*, ch. 3.

interesting feature of enhancement technologies is their connection with central aspects of a person's identity.[52] But what sense of identity is at issue?

One might think that numerical identity, the metaphysical relationship that an entity has to itself over time, is at issue. After developing a view of our numerical identity (and noting no other relevant sense of identity), Walter Glannon expresses this conviction in the context of genetic therapy:

> [G]ene therapy designed to correct or treat a cognitive or affective disorder would be more likely [than gene therapy with no direct effects on mental life] to alter one's identity. The manipulation of the relevant neurotransmitters or regions of the brain that generate and support mental life would directly affect the very nature of the mental states definitive of personhood and personal identity through time.[53]

Glannon applies this thesis to already existing persons, such as those with Down's syndrome, and not merely to embryos and newborns.[54] Presumably, his thesis would extend to enhancements that have comparably far-reaching consequences for someone's mental life.

But the thesis that improving a person's mental or psychological capabilities would create a numerically distinct individual is hardly credible. If the biological view of identity is correct, as argued in Chapter 2, the thesis is a nonstarter, because a single human organism survives such enhancements of mental life. More importantly, the thesis is incredible on *any* contending view of identity. For the person who exists after the enhancement will *remember* life before the intervention.[55] Moreover, the bulk of his intentions and attitudes are likely to survive the enhancement (though some may change, as they may in any one of us). Thus, even if one assumes with Glannon that we are essentially persons, there is no reason to think that enhancing an existing person's mental abilities would affect numerical identity. Although the psychological approach, as discussed in Chapter 2, understands our identity in terms of mental life, the criterion for identity is roughly persistence of one and the same mental life, not persistence of a mental life *with the same capacities and traits*. A plausible explanation of Glannon's reasoning is that he intuitively

---

[52] Elliott, *Better Than Well*, p. 26.

[53] *Genes and Future People* (Boulder, CO: Westview, 2001), pp. 81–2.

[54] Ibid., p. 90.

[55] More precisely, he will have detailed memory impressions of life before the intervention, and there is no good reason to deem these systematically erroneous (as they would be if numerical identity had been disrupted) for suggesting that he is remembering *his own* experiences. A similar qualification applies to the next sentence.

understands that a major change in capacities can affect *narrative* identity but, failing to distinguish numerical and narrative identity, he fallaciously applies this intuition to numerical identity.[56]

In any case, the original intuition is correct: Major changes in capacities or other traits can affect narrative identity. Recall that narrative identity involves our self-conceptions or implicit autobiographies. It concerns who we are – not in the sense relevant in a case of radical amnesia, but in the more everyday sense in which we ask what is most important in our lives: our self-told history, our core values and most important relationships. Because it involves our implicit autobiography not only up to the present, but also how we want the story to go in the future, it is identity in this sense with which adolescents as well as adults experiencing midlife crises struggle in trying to figure out who they are and where they want to go.

Note the place of enhancement projects in the narrative identities and self-creation of the individuals featured at the beginning of this chapter. Alan is mostly satisfied with his lot, having a strong family life and a good career, but turning forty has made him more sensitive about the little signs of retreating youthfulness. He feels middle-aged and mediocre. His modest enhancement project calls for restoring his hair color and getting in shape. What he wants in life determines where he wants his narrative to go, and at least for the next chapter of his story he wants these changes in himself. Because he consciously and deliberately pursues these changes, which he considers important, they constitute a self-creation project (as discussed in Chapter 3).

The desired enhancements in the other vignettes also constitute self-creation projects. Barbara's end is to become a more gaze-arresting young woman, her means cosmetic surgery and exercise. Carl plans to use an SSRI to change his personality and mental life in certain ways; he wants to become a person who is more confident and socially adept and less worried. In our futuristic vignette, Delia plans to employ genetic means to become a more awesome scholar. In deciding on these self-creation projects, our characters take a step toward defining themselves in the narrative sense, inasmuch as who someone is and where she wants her life to go are inextricably connected.

Are enhancement technologies inherently problematic for reasons connected with identity? We will consider in detail two major identity-related concerns.

---

[56] A similar conflation of numerical and narrative identity appears to muddy the conceptual waters in a discussion in Buchanan et al., *From Chance to Choice*, pp. 84–6.

THE CHARGE OF INAUTHENTICITY AND A POSSIBLE
BASIS IN EQUIVOCATION

Drawing from themes developed by Alasdair MacIntyre and Charles Taylor,[57] Carl Elliott has lately given voice to concerns about authenticity in relation to enhancement technologies. Elliott examines two competing ideals in American culture: (1) self-creation, with few if any constraints on what we can become, and (2) authenticity, which may imply some constraints. Understanding authenticity along the lines of Taylor, who characterizes it as being true to oneself, Elliott echoes a theme from Mac-Intyre in the following remark about this ideal: "We Americans talk about self-discovery, about finding ourselves, being true to who we really are. It is a kind of bad faith to pretend to be something that you're not – to discard your Jewishness, for example, or to move to the city and forget the folks back on the farm."[58] But the ideal of self-creation is also powerful, so that "much of American life is about this kind of struggle, between trying to reconcile yourself to who you are, on the one hand, and trying to change it on the other."[59]

What exactly is the concern about enhancement technologies in relation to authenticity? Interestingly, as Elliott notes, some individuals perceive enhancement technologies as *facilitating* their achievement of a more authentic self in projects of self-creation.[60] For example, after taking Prozac for several months before going off the medication, some patients have commented that only on the medication did they feel fully like themselves.[61] But the general tenor of Elliott's comments suggests the judgment that at least some uses of enhancement technologies are inauthentic, as expressed in this passage:

What is worrying about so-called "enhancement technologies" may not be the prospect of improvement but the more basic fact of altering oneself, of changing capacities and characteristics fundamental to one's identity.... [Deep] questions seem to be at issue when we talk about changing a person's identity, the very core of what the person is. Making him smarter, giving him a different personality or

---

[57] See MacIntyre, *After Virtue*, 2nd ed. (Notre Dame: University of Notre Dame Press, 1984), ch. 15; and Taylor, *Sources of the Self: The Making of Modern Identity* (Cambridge, MA: Harvard University Press, 1989) and *The Ethics of Authenticity* (Cambridge, MA: Harvard University Press, 1992).

[58] *Bioethics, Culture and Identity* (New York: Routledge, 1999), p. 33.

[59] Ibid.

[60] *Better Than Well*, ch. 2.

[61] Kramer, *Listening to Prozac*, ch. 1.

even giving him a new face – these things cut much closer to the bone.... They mean, in some sense, transforming him into a new person.[62]

Whether or not (as I conjecture) Elliott ultimately endorses these concerns, he gives expression to a type of worry that many people are likely to share. So does the President's Council on Bioethics: "In seeking by these [biotechnological] means to be better than we are or to like ourselves better than we do, we risk 'turning into someone else,' confounding the identity we have acquired through natural gift cultivated by genuinely lived experiences."[63] The concern is that enhancement technologies threaten to alter the self in some fundamental way, thereby changing a person's identity, transforming her into a new person – and that such change is objectionable for being inauthentic.

Once again, it is important to clarify that the sense of identity at issue is narrative identity. If taking an SSRI changes Carl's personality in important ways, *he* will change; it's not the case that one person would literally be destroyed and replaced with another, as would occur if numerical identity were disrupted. Now that it is clear that narrative identity is at issue, what exactly is the basis for concern?

Much of the concern about creating a "new person" may derive, once again, from conflating the two senses of identity, as in this reconstructed reasoning:

(1) Enhancement technology *E* alters a person's identity.
(2) Altering a person's identity is highly problematic. Therefore
(3) Enhancement technology *E* is highly problematic.

The fallacy should be clear. For premise (1) is true only if it appeals to narrative identity, while (2) is safely assumed only if it appeals to numerical identity. Equivocation on the term *identity* invalidates the inference to conclusion (3).

THE CHARGE OF VIOLATING INVIOLABLE
CORE CHARACTERISTICS

On an alternative reading, the reasoning previously reconstructed does not equivocate because identity is consistently understood in the narrative sense. Premise (1) remains true, but why accept (2)? What is wrong with changing someone's narrative identity, or self-conception, assuming

---

[62] *A Philosophical Disease*, pp. 28–9.
[63] *Beyond Therapy*, p. 300.

she autonomously consents to the change? In the passage quoted earlier, Elliott refers to "capacities and characteristics fundamental to one's identity" and "the very core of what the person is." Somewhat similarly, in a section entitled "Hubris or Humility: Respect for 'the Given'," the President's Council on Bioethics speaks of "a *human* 'givenness,' or a given humanness, that is also good and worth respecting."[64] Elliott and the Council never clarify what exactly they mean by these suggestive terms. Perhaps the most sympathetic interpretation is that some of our traits are so basic to who we are that they represent a sort of "core" of our narrative identities; changing parts of the core alters one's self-story so drastically that in a sense (that of narrative identity) the result is a different person. This charge of violating an inviolable core may be understood either as an elaboration of the charge of inauthenticity – to violate one's core is to fail to be true to oneself – or as a distinct moral objection.

In either case, why would it be problematic to transform someone into a new person by changing some trait at her core? (If the answer is "To do so would be inauthentic," naturally the next question is "How so?") For that matter, how do we determine which traits constitute this supposedly inviolable core? Perhaps the President's Council on Bioethics would answer "those that are 'given'" or "those that constitute human nature," but such responses would raise new questions about how to identify traits that are given or tied to human nature.[65] Furthermore, are traits that people would like to alter with enhancement technologies plausible candidates for inviolable core characteristics? I doubt that the present concern about enhancement technologies will survive the satisfactory answering of these questions.

First, it is hardly obvious that changing core traits is problematic if one autonomously consents to their alteration. The idea that some of our traits are inviolable strikes me as beholden to the rather implausible romantic notion of a "true" self whose defining traits are independent

---

[64] Ibid., p. 289. The Council also refers frequently to "human nature" and to "what it means to be a human being" (see, e.g., pp. 7–8).

[65] Francis Fukuyama contends that "the most significant threat posed by contemporary biotechnology is the possibility that it will alter human nature and thereby move us into a 'posthuman' stage of history" (*Our Posthuman Future* [New York: Picador, 2002], p. 7). He explains human nature in terms of allegedly unique cognitive faculties (ibid., pp. 140–3), but I was unable to find an argument as to why it would be problematic to alter these faculties. If, as he thinks, linguistic capacity is part of human nature, why would a genetic intervention that modestly improved someone's linguistic capacity be problematic?

of the individual's choice.[66] This notion of a true self seems most intelligible if construed as a person's *essence*, which is indeed independent of someone's choice, but the concept of essence is connected with numerical identity, which is not the sense of identity relevant to enhancement technologies.

But let's assume, at least for the sake of argument, that each person has some sort of narrative core. Which traits would belong to it? Presumably very basic characteristics associated with mental, social, and physical functioning – such as one's capacities to think, plan, care for others, walk, talk, and eat. And, then, for each of us there may be more personal or self-defining traits that we consider part of our narrative core: being an intellectual, say, or being humane. But it is one thing to claim that some trait lies within one's narrative core, another to claim that it is impermissible to *alter* that trait – that the trait is *inviolable*. I will argue that characteristics *likely to be targeted by enhancement technologies* are not plausibly regarded as inviolable. Let us begin with the traits highlighted in our vignettes.

Presumably, neither Alan's current level of physical fitness, how he feels when he turns turns forty, nor his semigray hair lies within an inviolable core. Barbara wants surgeons to cut into her body, possibly getting closer to her core. While breast augmentation raises significant prudential, moral, and social concerns (as discussed in the previous section), it does not seem to threaten Barbara's true self. To see this, consider accidental paths to breast enlargement (say, gaining a lot of weight) or deliberate change that is similar in kind but without cosmetic intent (e.g., breast reduction to ease pressure on the back). Surely these changes are not threatening to someone's core.

The concern may be more credible if qualified in this way: It is highly problematic to alter an individual's core traits *for frivolous reasons*.[67] One might agree that accidentally altering one's breasts, or intentionally doing so to improve one's health, is beyond reproach, while holding that intentionally doing so for something as frivolous as cosmetic improvement is indefensible or at least highly problematic. But Barbara's reasons for seeking these changes are not frivolous. Her desire to improve her dispiriting romantic life is surely a substantial, legitimate interest. The ambitions of the other three characters from our vignettes are similarly

---

[66] One might contend that appeals to authenticity are similarly beholden to this implausible notion. But I would disagree insofar as the demands of authenticity resolve into familiar demands of honesty and autonomy (see Chapter 3).

[67] In discussion, Maggie Little suggested roughly this amendment.

legitimate, as would be many people's projects of self-creation via enhancement technologies. While some desires for major self-change may be frivolously motivated – think of a contrarian teenager's desire to have a long surgical scar merely to impress fellow rebels – this possibility does not justify preventing, or even discouraging, major self-transformation through enhancement technologies *across the board*. (It does, however, provide ample reason for individual practitioners, such as the surgeon approached by the scar-seeking youth, to decline certain requests for services.)

Back to the vignettes. Carl's and Delia's enhancement projects offer somewhat better candidates for core characteristics: personality and internal psychological style in the case of Carl, memory and the biologically encoded need to sleep in Delia's case. Elliott mentions intelligence, which includes memory, and personality. Let us therefore consider several candidates for core traits, including some mentioned earlier that seem especially likely to be considered off limits if any traits are:

(1) Internal psychological style.
(2) Personality.
(3) General intelligence, including memory.
(4) The need to sleep a certain amount of time.
(5) Normal aging.
(6) Gender.

For each of these traits, our question is whether it is plausible to maintain that, morally, we should not deliberately change the trait in any significant way. Again, I do not deny that it might be wrong or morally troubling to change an allegedly core trait for truly frivolous reasons – say, having a sex-change operation on a whim. My concern is the bolder claim that certain traits are inviolable, such that they should not be altered even in pursuit of important life objectives.

First, *internal psychological style* – for example, depressed versus upbeat, worried versus confident, suspicious versus trusting, and all gradations between these poles – is a characteristic we may faultlessly change. Indeed, most psychotherapy has such self-transformation among its aims. In cases of enhancement, of course, an individual wants to improve her internal life beyond what is needed to qualify as healthy. But there is no reason to think that such an aim violates an untouchable core, since there is no basis for thinking that a particular psychological style is obligatory for a particular person.

Admittedly, truly extreme changes in psychology that cut one off from reality would be problematic. If we could have a very pleasant life only at the cost of profound, systematic misunderstanding of reality, most of us would not consider that better for us than a less pleasant reality-based life.[68] But while the possibility of hedonistic bliss while one is completely cut off from reality is important to science fiction and philosophy, it is not a significant concern in this discussion. To be sure, the sorts of changes in psychological experience we are considering might affect one's perception of social reality – say, by affecting mood – but there is no reason to think that the individual's resulting perceptions would be unreasonable. And, even if in some rare cases an enhancement technology distorted perceptions alarmingly, this would in no way suggest that internal psychological style was an inviolable core characteristic. It would simply suggest that some very drastic changes to this characteristic would be problematic and perhaps unacceptable.

Is *personality* a more plausible candidate for an untouchable characteristic? Elliott suggests its plausibility, and I have noticed in conversation that confirming opinions are common. But, on reflection, personality appears no more promising a candidate than internal psychological style. When someone has an unattractive or self-defeating personality – say, overly cynical and sarcastic, or buffoonish, or excessively shy – we generally applaud his efforts to make positive changes. Admittedly, some personalities that are not considered the most attractive are closely linked with moral virtues and for that reason might be better left untouched; for example, a modest and very sincere individual might not sparkle at cocktail parties. But the close association in certain cases between virtues and (from a mainstream perspective) less than ideal personalities should not obscure the appropriateness of certain enhancements in personality. We should also acknowledge that truly extreme changes in personality may create problems for interpersonal relations, as associates may no longer know how to relate to the "new" person. But realistic scenarios do not involves such drastic changes, and their mere possibility does nothing to show that deliberately changing one's personality is inherently problematic. How to explain, then, the inclination to believe one's personality is untouchable? Perhaps the term *personality* reminds us of the term *person* – and the idea of changing the person creates vague

---

[68] This is roughly the point of Robert Nozick's experience machine thought experiment (*Anarchy, State, and Utopia* [New York: Basic Books, 1974], pp. 42–5). See also President's Council on Bioethics, *Beyond Therapy*, ch. 5.

worries about disrupting identity, stemming again from a failure to distinguish numerical and narrative identity.

How about *intelligence*: memory, analytical ability, social and introspective perceptiveness, imaginativeness, and so on? Well, attentive parents try to improve their children's intelligence all the time with stimulating home environments, preschool, educational games, and other mind builders. Nowadays retired persons are encouraged to keep mentally active by reading, playing a musical instrument, doing crossword, puzzles, and the like. And, between the early years and old age, many of us try to improve or maintain our intelligence by reading a lot, playing challenging games such as chess, and limiting television viewing and alcohol consumption; some believe the music one listens to also matters. No one objects to these enhancements of intelligence.

One might respond that my examples of innocuous intelligence promotion are not really examples of enhancement. Rather, they are measures that enable someone to reach or maintain her full level of intelligence, with which she is *born*. But, of course, prenatal nutrition and the presence or absence of noxious chemicals in the mother's bloodstream can greatly affect the fetus's later prospects for mental functioning. So the objection is more plausible if it appeals to the full level of intellectual potential as determined by one's original *genome*. But to take one's original genome as essential to one's "true" intelligence seems overly reductionistic for two reasons.

First, genes interact constantly with the environment to determine our capacities, other characteristics, and level of functioning. There seems no good reason to treat one's genome as more fundamental to one's intelligence than the interaction of one's genome with the environment.[69] Consider this analogy. My genome has always encoded the potential to learn languages. I grew up where mostly English is spoken, so I speak English fluently. Had I died as a fetus or newborn, or had I been lost and raised entirely by bears, I would not have spoken any language despite my genetic endowment. Both factors, my genes and my environment, seem crucial to my language-speaking ability. A corresponding judgment about intelligence is plausible. Your fine genome notwithstanding, had your diet and other environmental factors made you profoundly mentally disabled, you would not have been intelligent; that is, you would not have had the capacities that constitute intelligence. Intelligence involves certain *capacities* (which, before adulthood, are relativized to age), not

[69] See Buchanan et al., *From Chance to Choice*, p. 160.

mere *potential*, which is why it would be nonsense to say that someone had been an intelligent fetus.

Second, our genome changes over time, as mutations spontaneously occur, so to favor our *original* genome over all the later versions would be highly arbitrary. McGee puts it well:

> [W]hat looks like a stable matrix – this "genome" we're all trying to protect – is actually a complex matrix of interactions in three trillion cells, many of which have taken up small mutations. Walking in the sun ages your skin because it affects your genetic makeup. Radiation and the chemicals in water effect changes in the germ-line and somatic cells. Even the air that you breathe is chock-full of ingredients that change the supposedly stable "blueprint" of genetics.[70]

Even if our genome deserved priority over our environment and counted as part of our inalienable core, it would make more sense to treat our *present* genome as crucial. After all, if some lucky mutation greatly increased my intelligence, presumably the new intellectual baseline would count as more relevant to my core than my original baseline. Inasmuch as genetic enhancement of intelligence would give me a new baseline, it would not violate my core; it would reconstitute it.

One might object that genetic enhancement of intelligence, unlike the lucky mutation, violates one's core because the former is a deliberate product of human agency. "Mother Nature may mess with Mother Nature," one might say, "but *we* shouldn't mess with Mother Nature – or at least certain parts of her, like the genes that code for intellectual capability." Well, why not? Note that the objection still assumes, implausibly, that one's genome is fundamental in a way that environmental factors are not. They are equally important, so we should remember that no one objects to deliberately introducing environmental factors that promote intelligence.

At this point, my opponent could take a different tack: "I agree that the aim of improving someone's intelligence is not objectionable per se. What is objectionable is the *means* of deliberately changing one's genome." But if genes and environment interact constantly to determine one's intellectual ability, and if deliberately manipulating the environment to improve intelligence is beyond reproach, why should a different, possibly more efficient means to the same laudable end be problematic? My opponent might again appeal to the idea of a core, claiming that genetic interventions alter our core, whereas environmental influences do not. But

[70] *The Perfect Baby*, p. 118.

this is false, for both genetic and environmental interventions affect, for example, the properties of our brain.

I conclude, then, that the case for regarding intelligence as a core characteristic off limits to enhancement technologies is unpromising. But *the need to sleep a certain amount of time*, which is equally a function of brain structure and functioning, seems no more promising as a candidate for an inviolable characteristic. One's need for sleep can be modified in ordinary, unobjectionable ways. Some people find that vigorous exercise such as swimming or deep relaxation techniques such as meditation can modestly reduce their need for sleep; no one objects to these means. As in the case of intelligence, moreover, the fact that genetic interventions work directly on the genome is unimportant. Nothing about the need to sleep makes it a better candidate than intelligence for an inviolable trait.

Is *normal aging* a better candidate? Whereas genetic alterations of intelligence or the need to sleep aim at *changing the level or degree* of a universal human characteristic, genetic interventions to prevent normal aging would aim to *eliminate* what we have heretofore regarded as a universal human characteristic. Aging, one might argue, is an essential part of any recognizable human life. We should therefore not permit attempts to enhance individuals by preventing them from aging through genetic means.

This argument assumes that the elimination of aging is possible in principle. So, in claiming that aging is *essential* to human life, it presumably understands this term in the normative sense of inviolability. One difficulty of assessing this normative claim is the difficulty of imaging a human life in which one *never* aged – in the familiar sense of aging that involves gradual bodily deterioration. (Of course, it's *contradictory* to suppose that someone could stay alive and never age in the sense of acquiring more birthdays.) If one never deteriorated and avoided sudden causes of death, one could live for hundreds of years – though presumably at some point the Grim Reaper would catch up with one with a fatal accident, murder, or perhaps boredom-inspired suicide. Although I do not find the prospect of living 500 or 1,000 years attractive, others might. If there proves to be a demand for the development of an antiaging genetic technique, then those seeking it may be voicing their view that aging is not normatively necessary. But, admittedly, I'm not sure how to decide the issue of whether aging is an inviolable characteristic.

But even if we agreed that aging is inviolable, that would hardly cast doubt on efforts to *slow or delay* aging. Understood thus, antiaging genetics would seem to be of a piece with other technologies that change the degree or specific expression of some human (and animal) phenomenon such as intelligence or the need to sleep. More broadly, such a technology would be of a piece with mainstream medicine and the field of public health, which attempt in innumerable ways to postpone our deaths.[71] In the twentieth century, the life expectancy for Americans increased about thirty years, due in large part to advances in these two fields.[72] So even if aging is an inviolable core trait of human beings, living no longer than some number of years is not. Thus the prospect of genetic interventions to delay aging remains in good standing.

Is *gender* a better candidate for an inviolable core characteristic? Probably most of us take our gender as a central part of our self-conceptions, as central to our narrative identities. One might even hold that someone's gender is essential in the strict metaphysical sense that a person is essentially of a particular gender, so that no one of a different gender at another time could be numerically identical to that person. The President's Council on Bioethics suggests this stronger essentialist claim: "[W]e will have to take sexual identity seriously as given with our body. Every cell of the body marks us as either male or female, and it is hard to imagine any more fundamental or essential characteristic of a person."[73]

But, in view of the experiences of some transgendered persons, the essentialist thesis is very doubtful. Someone who grew up in my neighborhood had a sex-change operation in his thirties. For many years he was male both biologically and psychologically – that is, in terms of his self-conception. At some point in late adolescence or early adulthood, Chris began to regard himself as "really" female. Years later he became certain of this transformation, had a supporting surgery, and socially became "Christine." Well, Christine remembers being male. By all reasonable criteria, she was, years ago, a person of the male gender. But she changed, becoming female.

There are two main ways to counter my assertion that Chris(tine) changed genders, but neither is promising. One might claim that Christine is still male – her self-conception, social persona, and current genitalia notwithstanding – because gender is a purely biological matter determined entirely by one's original genetic constitution. Now, certainly one can insist on using the terms *gender*, *male*, and *female* in this way, but

---

[71] Indeed, if advances in biology confirm the theory that aging is a not-inevitable susceptibility to illnesses (see footnote 26), an antiaging technology may need to be reclassified as treatment or prevention, as opposed to enhancement.

[72] President's Council on Bioethics, *Beyond Therapy*, p. 165.

[73] Ibid., p. 69.

this usage does not capture the more nuanced sense of these terms that is customary today. I am using these terms such that it is conceptually possible to change genders. Another way to deny that someone could change genders appeals to a criterion of numerical identity according to which one's original gender is essential in the strictest sense, as the President's Council suggests. This implies that Christine's "memories" are delusional, that it was not one and the same human individual who had been male, but rather a predecessor who was replaced – without dying? – by a successor who wrongly takes her presurgery "memories" to be self-representing. This implication is extraordinarily implausible. It is far more sensible to allow that someone who was male became female.

Cases like this show that, while gender may be crucial to someone's narrative identity *at a particular time*, gender is not criterial to numerical identity. And because one can change one's self-conception of gender, the latter can be dynamic with respect to narrative identity. Gender is not an inviolable core characteristic, so neither genetic nor surgical interventions that would change a person's gender are cogently criticized on this ground.

In conclusion, the thesis that certain core characteristics *likely to be changed through enhancement technologies* are inviolable is highly dubious. Even if aging, for example, is a promising candidate for an inviolable core trait, enhancement technologies are unlikely to target aging itself. Rather, they would target one's *current expected life span*, which is hardly inviolable.

Those who understand the inviolability thesis as an elaboration of the charge of inauthenticity may state our conclusion this way: Even the changing of core characteristics through enhancement technologies can be perfectly authentic expressions of an agent's values in self-creation. Authenticity has little or nothing to do with the *degree* of self-change sought and achieved. Rather, as argued in Chapter 3, it is a function of the autonomy and honesty of the agent's self-creation project.

It bears repeating, however, that this critique of the violability thesis does not entail the judgment that any and all efforts to change core traits, assuming there are core traits, are beyond reproach. I remain open to the possibility that it is wrong, or at least highly problematic, to change core traits *for frivolous reasons* – for example, seeking to make oneself less intelligent through neurosurgery just because one thinks this would be funny. My speculation is that if actions of this sort are morally problematic, that is because they demonstrate insufficient self-respect. Naturally, an adequate defense of this position would require further argument that the class of actions in question is morally problematic, that we have obligations grounded in self-respect, and that such obligations provide the basis for criticizing this class of actions.[74] Being merely open, rather than committed to, this thesis and the suggested grounding in self-respect, I will not pursue this issue further here.

CONCLUSION

Enhancement technologies are not new. Hair dye, steroids, and facelifts, for example, have been available for decades. Moreover, the use of a particular product or procedure may count as an enhancement in one health context and as medical treatment in another, as with steroids and nasal reconstruction, reminding us that the technologies themselves need not be novel. At the same time, some specific types of enhancement technologies are new or even unrealized. Hence the forward-leaning feel of discussions of SSRIs and genetic enhancements. Also relatively new is scholarship on enhancements as such.

In this chapter, we examined various moral and social issues provoked by enhancement technologies. Some of these concerns were relatively familiar. As a whole, they highlighted areas of legitimate concern that should be addressed case by case as we examine particular enhancement technologies (e.g., a genetic technology aimed at improving memory), or specific cases of individuals seeking an enhancement or making it professionally available (e.g., a surgeon advertising facelifts in a particular way). Concerns about justice arguably supported universal access to health care as a moral prerequisite, in any relatively wealthy country that has not achieved this goal, for further development of biomedical enhancements. Other, less familiar issues concerned human identity. We found that reflection on identity neither ruled out the use of enhancement technologies on grounds of inauthenticity nor vindicated the judgment that certain core traits likely to be targeted by enhancement technologies are inviolable and therefore off limits to deliberate change.

---

[74] Would such a requirement of self-respect be tied to that of authenticity? Perhaps authenticity requires not only autonomy and honesty, but also a measure of self-respect. Although I suspect not, believing that someone might authentically show major disrespect for himself, I realize that some might want to specify "being true to oneself" in a way that demands self-respect. One might even argue that self-respect is required for autonomy, one of the conditions of authenticity; about this, too, I am skeptical, though I will not pursue the matter here.

Our discussion left open the possibility that some enhancement technologies, or particular uses of them, should be avoided, discouraged, or prohibited – even in countries with universal access to health care – either because they present very poor risk–benefit ratios or because a particular use would be frivolous and insufficiently self-respecting. Our reflections did not support the judgment that some general categories of enhancement, such as cosmetic surgery, were beyond the moral and policy pale. We conclude, therefore, on a note of cautious openness to the use of enhancement technologies in projects of self-creation.

7

Prenatal Identity

Genetic Interventions, Reproductive Choices

According to the framework developed in this book, we human persons are essentially human animals and characteristically self-narrators who care about continuing as such. Through the lens of this framework, in conjunction with defensible moral assumptions, we have examined the definition of human death, the authority of advance directives in cases involving severe dementia, and the use of enhancement technologies. Our discussion of enhancement technologies focused on *postnatal* enhancements, while our treatment of advance directives considered an institution in which adults make self-regarding decisions. Naturally, different issues arise when adults make decisions for dependent others – with greater complexity emerging if those others are unborn. Our examination of the definition of death studied the end of our existence, leaving open the question of when we come into existence.

In this final chapter, we will confront several controversial issues that were not addressed in earlier chapters. We will consider the prenatal human animal, or organism, and several types of decisions that adults make on behalf of this being. More specifically, we will address these topics: (1) our origins, or when we come into existence; (2) prenatal genetic interventions in relation to human identity; (3) the ethics of prenatal genetic interventions; (4) reproductive choices that may affect who comes into being (the nonidentity problem); and (5) implications of this chapter's findings and the Time-Relative Interest Account (TRIA) (introduced in Chapter 5) for the morality of abortion.

Before addressing our origins, let us note a point that is often overlooked: When we come into existence entails nothing about our moral status once we exist. I will argue that we originate early in gestation.

As discussed in the final section, this view of our origins *leaves open the possibility* that the prenatal human has significant moral status early in pregnancy – possibly, as soon as it comes into being. But there is no incoherence in the notion that we might exist before acquiring significant moral status. Only moral argument, as I will provide, can determine whether that is the case.[1]

OUR ORIGINS

When did we come into existence? At what point in prenatal development did the being that is now one of us first appear? Since the question concerns our existence conditions, its answer depends on a correct account of numerical identity. According to our account, the biological view, we are essentially human animals. So our question is when the human animal comes into existence. Or, since the word *organism* sounds more natural than *animal* in referring to the developing creature early in pregnancy, we may ask when the human organism emerges.[2]

This question is logically related to the issue of when the human organism goes out of existence or dies. The question of human death – at least when the issue is construed ontologically as opposed to pragmatically – asks when one of us literally goes out of existence (assuming there is no afterlife[3]). Death occurs when life ends. And life – in the literal, original biological sense – is distinguished from inanimate matter in roughly this way: *Living things are complex products of reproduction – from similar parents, through evolution – that use energy from their environment to impose or maintain internal organization and resist entropy.*[4] A corpse, by contrast, does not use

---

[1] Alfonso Gomez-Lobo argues, to the contrary, that we must acknowledge that we have full moral status – he speaks of *inviolability* – from the time we come into being; otherwise, one is in the position of judging that it "would have been morally permissible to destroy me in the past so that my later inviolability would be worthless" ("On the Ethical Evaluation of Stem Cell Research: Remarks on a Paper by N. Knoepffler," *Kennedy Institute of Ethics Journal* 14 [2004], p. 79). Perceiving no absurdity in the implication that it would have been permissible to destroy me in utero, and not finding self-evident the thesis that we have full moral status from the time of our origination, I hold that when we acquire moral status is an open question.

[2] Lee Silver informed me that biologists typically favor the term *organism* over *animal* in referring to the prenatal being until certain species-typical features develop. He clarified that his point was purely semantic and did not imply that a new being, the animal, replaces the original organism.

[3] In Chapter 2, I briefly defended this assumption. Those unable to accept it may substitute the qualification "at least in this world" for "assuming there is no afterlife."

[4] Cf. Lee Silver, *Remaking Eden* (New York: Avon, 1997), ch. 1.

energy from the environment to sustain internal organization; its organs lack the integrated functioning characteristic of life, so the body tends to succumb to entropy and decay.

One might think that the conceptual connection between our origins and our death threatens the biological view of human identity. After all, a typical human being near death is thoroughly unlike an early fetus. So how can the biological view, which presumably claims that the two beings are one and the same, be correct? Well, the two beings *are* similar in one crucial respect: They are both (living) human organisms. So we return to the question of when the human organism that is one of us emerges in gestation.

Some people who accept the biological view assume that we come into existence at conception, as single-cell zygotes. (I will use the term *zygote*, more broadly than it is sometimes used, to refer to the developing entity until implantation into the uterine wall.) Conservatives on the morality of abortion typically make this assumption, as does one of their most influential voices, the Vatican.[5] In light of certain facts about prenatal development, I will argue that this assumption proves indefensible.

After raising the issue of whether a particular person, such as Queen Elizabeth, might have had different parents in "another possible world" (a counterfactual situation), Saul Kripke answers with a rhetorical question: "How could a person originating from different parents, from a totally different sperm and egg, be *this very woman?*"[6] Kripke is suggesting that a given individual could not have come into being, in a counterfactual situation, from different parents – or even from a different sperm and egg – than those from whom she did in fact derive. I agree. You would never have come into being if the very sperm and egg from which you in fact derived had never joined in fertilization. But from this it does not follow that you originated as that zygote, the immediate product of fertilization. For while its existence was necessary for your existence – in this, the actual world, or any other possible world – maybe it wasn't sufficient. I am suggesting that it was not sufficient for your existence and that the zygote was a sort of precursor to you.

A single-cell zygote is not a uniquely individuated human organism – of the sort that each of us is. Indeed, there is not a uniquely individuated human

---

[5] See, e.g., Pope John Paul II, "The Unspeakable Crime of Abortion," in *Evangelium Vitae*, encyclical letter of John Paul II, March 25, 1995 (Vatican City: Libreria Editrice Vaticana, 1995).

[6] *Naming and Necessity*, 2nd ed. (Cambridge, MA: Harvard University Press, 1980), p. 112.

organism, of the sort that we are, until *at least* several days after conception. To see why, consider some basic facts about human reproduction.

For about two weeks following fertilization, the zygote (leaving aside for now whether it is a human organism of the sort that we are) is, in an important sense, not uniquely individuated. That is because it can still divide into two or more parts that go on to develop into human beings, and it can still merge with another zygote and develop into a single human being.

Let us refer to the case of division as *monozygotic twinning* even though division into triplets or even quadruplets, while rare, is also possible. The resulting zygotes, from which identical twins derive, have virtually identical DNA.[7,8] By contrast, *fraternal* twins, resulting from the fertilization of two eggs in the same cycle, are likely to have no more genetic similarity than ordinary siblings. Let us term *fusion* that unusual occurrence in which two zygotes, the result of fraternal twinning, merge into one zygote, a chimera. The chimera has two complete sets of DNA, which somehow compete, or perhaps cooperate, in determining (along with the prenatal and postnatal environments) the individual's phenotype – his actual characteristics, such as eye color, height, and talents. Probably most chimeras go through life without any awareness of having two complete sets of DNA. But sometimes this fact comes to light – for example, when doctors investigating problems with someone's reproductive system find that the patient has both male and female sex organs, or when blood tests suggest that a child's biological mother "could not possibly be" his biological mother, or when the girl down the street has one brown and one hazel eye.[9]

So, until about two weeks after fertilization, a zygote can divide into two zygotes, and in cases of fraternal twinning, two zygotes can fuse into one.

---

[7] I say "virtually" because mutations can introduce minor differences between the two zygotes' DNA, as can mitochondrial DNA in the cytoplasm surrounding the nucleus. Each nucleus, of course, contains the forty-six chromosomes that are replicated in monozygotic twinning.

[8] Although monozygotic twinning occurs naturally, it can also be induced artificially in what is called *embryo splitting*. In this procedure, scientists wait until the zygote has divided to the four- or eight-cell stage, then remove the zona pellucida – the thin membrane that surrounds the cells, keeping them together – before teasing apart the individual cells and encapsulating each with an artificial equivalent to the zona pellucida.

[9] See Claire Ainsworth, "The Stranger Within," *New Scientist* (November 15, 2003): 34–7. Although fusion occurs naturally, in principle it could also be achieved artificially by bringing two zygotes together and inducing them with an artificial zona to develop as one. To my knowledge, this has not been done.

It would therefore seem that the zygote is not yet uniquely individuated in the sense that whether it, and it alone, will develop into a single human organism has yet to be determined. It is not yet a unique member of our kind, *human organism*; it may split into more than one or it may combine with another to form an individual with a significantly different genome.

Consider any two adult identical twins and any adult chimera. Both twins derived from a single zygote. If each human organism originated as a one-cell zygote, then each of the two twins originated as *the same* zygote. But that is an incoherent result. For the two twins are numerically distinct, so they cannot *both* be identical to a single earlier zygote.[10] Meanwhile, no chimera could have originated from a single one-cell zygote because it took two distinct zygotes to furnish her genome. Surely she wasn't individuated until fusion took place; and the twins were not individuated until twinning occurred. Neither identical twins nor chimeras could have come into existence at conception.

One might wonder, however, whether human beings who are neither identical twins nor chimeras might have originated at conception. After all, those who are the unique continuants (unlike identical twins) of a single zygote (unlike chimeras) can trace a continual path of development from conception to the present. But the possibility of tracing this developmental path is consistent with my claim that the single-cell zygote is merely a precursor to a human organism like you or me. This claim is supported both by both counterfactual considerations and by further facts about prenatal development.

Consider the zygote my parents produced in 1961, leading to my birth in 1962. I am not an identical twin. But that zygote could have split spontaneously, resulting in identical twins. If it had, presumably I would not have existed, because it is implausible to identify me with either of the twins in that counterfactual scenario. If that is right, then the existence of the zygote my parents produced was not sufficient for my existence, from which it follows that I am not numerically identical to that zygote. The very possibility of twinning belies the claim that we originated at conception.

A rebuttal is possible, however. One might argue that, in the counterfactual twinning scenario, I did exist – very briefly, as a zygote – before

---

[10] Here I make the standard assumption that identity is transitive: If $A = B$ and $B = C$, then $A = C$.

I vanished at the time of twinning; every time a zygote splits in natural twinning, a human organism (of our kind) goes out of existence. This follows from the thesis that, while nontwins come into existence at the time of conception, twins come into existence at the time of twinning.[11]

But this thesis faces difficulties. First, it implies that anyone who believes that we have full moral status from the moment we originate has reason to mourn the loss of the original zygote in every case of twinning. Moreover, such a person has ample reason to support a substantial public investment in research that could lead to the prevention of twinning, just as we invest heavily in research that promotes our ability to forestall postnatal loss of life. I imagine, though, that among people holding this moral view, *very* few would accept these implications. And that, I suggest, is because the conclusion is false, as is the premise that a human organism (of our kind) comes into existence at conception and goes out of existence at twinning.

The preceding thesis has other difficulties as well. Surely it is a little odd that twins and the rest of us – who are of the same basic kind – should originate at different points in gestation. While this oddity by itself is no theoretical fatal flaw, it is connected with a more significant concern. Arguments for the claim that (nontwin) human beings originate at conception emphasize the importance of *continuity* from conception to any later stage. These arguments deny that the zygote could be a *precursor* to an individual of our kind (as I claim) because "[T]he predecessor of something... cannot be continuous with the entity of which it is the predecessor."[12] But there is spatiotemporal continuity between the zygote and the later twins, each of whom can trace her developmental path back to a zygote whose complete human genome established theirs. Now, again, it is impossible for both twins to be numerically identical to the zygote. I suggest, then, that in no cases of prenatal development – including those involving nontwins – are spatiotemporal continuity and continuation of the same complete genome sufficient to establish identity, for the product of conception is not yet uniquely individuated. None of us comes into existence as a single-cell zygote.

Additional facts about prenatal development further support this claim. So far, we have emphasized that the zygote is not uniquely individuated. But, as we will soon see, *the zygote is not even a human organism*

---

[11] See Gomez-Lobo, "On the Ethical Evaluation of Stem Cell Research," p. 79.
[12] Ibid., p. 78.

of the type that you and I are; for the first few cell divisions, it is more like a colony
of cells loosely joined by the membrane known as the zona pellucida.

Before turning to the facts that bear out this assertion, a few clarifica-
tions are in order. As stressed in Chapter 4, organisms are the basic sorts
of things that live and die. They are characterized by internal complexity
featuring the interdependence of different subsystems, and the taking of
energy from the environment to maintain internal order and resist entropy. But organisms come in enormous varieties. Some, such as amebas,
consist of only one cell. Even a cell features enormous internal complexity and interdependence of subsystems. Now, to be sure, the single-cell
human zygote is an organism. And it is a human organism as contrasted,
say, with a bovine or canine organism. But so is a sperm cell and so is
an unfertilized egg – a fact that should make us hesitate before infer-
ring from the fact that a zygote is a human organism that it is a human
organism of the same kind as you and I.

Although a zygote, unlike an egg or sperm cell, has the same number
of chromosomes as each of us does, after one cell division and for several
divisions thereafter it functions less like a single integrated, energy-using
entity – of the sort we call an organism – than like a collection of single-cell
organisms loosely and contingently stuck together. That is precisely why
twinning and fusion remain possible. (The very *nature* of the zygote entails
the possibility of twinning or fusion and, with this possibility, the absence
of unique individuation.) After the sperm cell enters the egg, their two
sets of twenty-three chromosomes remain separate for about a day. As
biologist Lee Silver explains, contrary to popular belief, fertilization itself
is completed only at the two-cell stage:

> [T]he chromosomes in the two pronuclei duplicate themselves separately, and
> then copies from each come together inside the actual nuclei formed *after* the
> first cell division. It is within each of the two nuclei present in the two-cell embryo
> that a complete set of forty-six human chromosomes commingle for the first
> time.[13]

Two further divisions in the next two days yield an eight-cell zygote. Importantly, each of the eight cells retains the capacity, if separated, to produce
a human organism like you or me.[14] So far, there is no specialization of
these cells to perform different tasks; nor is there interaction or integration among them. In this sense, they are tantamount to a colony of

---

[13] Silver, *Remaking Eden*, p. 45.
[14] Ibid., p. 58.

eight contingently joined zygotes. They are not yet functioning as a single organism.

Differentiation of cells begins at the sixteen-cell stage, when the outer cells begin to transform into what will become the placenta.[15] Division continues as the zygote moves through the fallopian tube, entering the uterus on about the fifth day after fertilization. (By *fertilization* I mean *the beginning of* the process of fertilization.) Between seven and eight days after fertilization, it attaches to and penetrates the uterine wall, making connections with the mother's blood supply. Now, for the first time the embryo begins to absorb nutrients, gaining energy from its environment rather than from internal reserves[16] – one of the characteristic features of an organism. Still, the middle cells have not differentiated and the embryo could split spontaneously into more than one viable organism. On day 14 or 15, some middle cells differentiate and it is now determined which cells will become part of the fetus and which will become part of the placenta. Within a day, in the portion that will become the fetus, a line of cells differentiates into the *primitive streak*, the precursor to the spinal cord. Natural or spontaneous twinning is now impossible.[17] The human organism is now uniquely individuated, and it clearly functions as a single integrated unit. From a biological understanding of what we are, therefore, there is no conceptual obstacle to identifying it as a being of our kind.

Thus, once spontaneous twinning has been precluded and all parts of the embryo are differentiated at about two weeks after fertilization, a human organism of the type that we are has emerged. Let us hereafter reserve the term *human organism* for those of this type, making the qualification "of the type we are" optional (despite excluding sperm cells, unfertilized eggs, and zygotes). But, arguably, we should allow that the human organism emerges earlier, perhaps when differentiation *begins* at the sixteen-cell stage – making the full package "more than the sum of its parts" – or when the embryo implants and begins to absorb energy from its external environment. I am less confident about drawing a fairly specific line marking the origins of the human organism than I am in

---

[15] In this paragraph I have benefited greatly from Silver, *Remaking Eden*, pp. 58–63.

[16] This particular point is made by Eric Olson, *The Human Animal* (New York: Oxford University Press, 1997), p. 91.

[17] In principle, artificial, delayed twinning via cloning remains possible. This would involve removing a somatic cell from the human organism, stimulating it to return to its undifferentiated state, and fusing it with an egg from which the nucleus has been removed to create a new zygote.

defending two claims: (1) We do not exist, before the sixteen-cell stage, as completely undifferentiated, preimplantation zygotes, and (2) by the time all parts of the embryo are differentiated and twinning is precluded, one of us has come into being. Perhaps we come into existence somewhat gradually, emerging over a nontrivial stretch of time rather than at a single moment.

Having argued that each of us came into existence between the sixteen-cell stage and full differentiation – that this is the most plausible specification of the biological view's implication regarding our origins – I would like to underscore the specific thesis that we did not come into existence at conception as single-cell zygotes. Recall that it is not until the two-cell stage that the genetic material from the sperm and egg make contact, creating a full complement of DNA. Furthermore, the almost universal acceptance of prenatal genetic diagnosis intuitively confirms that the single-cell zygote is not one of us.[18] In this practice, a single cell is removed from a preimplantation zygote in order to test it for predisposition to one or more genetically determined diseases such as Huntington's disease or cystic fibrosis. But, once this cell is removed, it is effectively an independent zygote; provided an artificial zona pellucida, it could develop like any other zygote. Yet not even the Catholic Church objects to this procedure, which invariably results in the death of the cell whose DNA is examined. This suggests an intuitive understanding that a full complement of human DNA is insufficient to constitute one of us.

Clearly, the single-cell zygote has the *potential* to develop in such a way that eventually produces one of us. (Note: I do not say that the single-cell zygote has the potential to *become* one of us – a statement that would imply numerical identity.) But the importance of this potential is dubious. Now that we know that mammals can be cloned from somatic cells – bodily cells other than sperm, eggs, and their stem-cell precursors – we know that, in principle, each of millions of cells in your body has the potential to develop into a full human organism. Surely this confers no particular moral status on your many individual cells; nor does it suggest that each cell is one of us. Once again, a full complement of DNA is not enough to make one of us.

One might reply that neither a cell extracted for prenatal genetic diagnosis nor an ordinary somatic cell is in a state that *naturally* leads to cell division and eventual birth. Perhaps natural potential matters in a

---

[18] Cf. Silver, *Remaking Eden*, p. 53.

way that artificial potential does not. (If so, then presumably we should not ascribe significant potential to spare embryos, left over from in vitro fertilization, which have not been implanted into a mother's uterus.) But neither is a single-cell zygote produced by ordinary fertilization in a state that naturally leads to birth – if by *natural* we mean "most likely to occur" – for *approximately 75 percent of pregnancies terminate spontaneously.*[19] To be sure, some will understand potential by reference to a different sense of *natural,* such that the natural course of events in pregnancy leads to birth. I do not doubt that some such sense of *natural* is meaningful. Nevertheless, if the single-cell zygote is one of us, then two extraordinary implications follow: (1) Three-fourths of humanity is never born and (2) those who attribute full moral status to us from the first moment of our existence ought to grieve for all zygotes lost in spontaneous abortions, including the majority of them that occur prior to implantation without the mother's even knowing she was pregnant. (And those who believe in an afterlife must picture unborn beings constituting its majority population.) I suspect that nearly everyone will find these implications incredible. With this point, I complete my argument that we do not originate prior to the sixteen-cell stage, much less as single-cell zygotes.

We come into existence some time between the sixteen-cell stage and the time at which differentiation characterizes all portions of the embryo and twinning becomes impossible. Let's call this time (without being sure of exactly when it is) the *time of unique individuation.* Now, anything that exists – that has come into being – can either continue to exist through change or go out of existence due to change. *The question of prenatal identity over time through change* asks what criteria determine whether a prenatal human organism persists through change. Now, for anything that exists, certain conditions had to obtain for it to have come into being. So certain counterfactual situations were compatible with its origination, while other counterfactual situations would have precluded its origination and perhaps ensured the origination of a different individual. This raises *the question of preindividuation change in relation to later identity:* Which sorts of preindividuation change are likely to be identity-affecting in this sense? Genetic interventions that occur prior to the time of unique individuation provoke this question. Genetic interventions that occur after that point (but prenatally) provoke the question

---

[19] Ibid., pp. 50–1.

of prenatal identity over time through change. So, regardless of when prenatal genetic interventions occur, they raise issues regarding human identity.

PRENATAL GENETIC INTERVENTIONS AND HUMAN IDENTITY

Clarification of these issues concerning prenatal identity will prove important to our later discussion of the ethics of prenatal genetic interventions. As we will see, several commentators believe that genetic therapy and/or enhancement would, in certain cases, entail a change of numerical identity. In these cases, either (1) the intervention results in the origination of a different individual than would otherwise have originated or (2) the intervention puts one individual out of existence and creates another in its wake. If these commentators are right, then what we ordinarily think of as genetic therapy or enhancement may – by preventing the existence of, or eliminating, one prenatal individual and creating another with a slightly preferable genome – amount to a form of eugenics! But are they right? How much genetic change is compatible with a given human organism's origination or continued existence?

This question returns us to the topic of *individual essentialism,* as briefly discussed in Chapter 4. To say that I am essentially a human organism is to say that I am essentially a member of a particular kind: *human organism.* But I am not just any old human organism; I am *this* one. You, too, are essentially a human organism, yet you and I and everyone else are essentially distinct from each other. This helps to explain what's funny about an old joke: "The other day I was walking down . . . oh, no, that must have been someone else." But, while the distinction between a self-aware person and anyone else may be laughingly obvious, the distinction between one prenatal human organism and another that imperceptibly replaces it is not.

Let's consider different stages at which prenatal genetic interventions may occur. Since these interventions are not yet technically feasible, our discussion will concern future possibilities rather than current practice. In principle it will be possible to perform a genetic intervention either on human gametes (sperm or egg cells) prior to conception or on the product of conception. If the latter, genetic interventions may occur either prior to or after unique individuation. Wherever the line should be drawn indicating the time one of us originates – a line that will have some breadth if we come into existence gradually – there is an important metaphysical and conceptual difference between interventions before and those after that time. Let me explain, beginning with preindividuation interventions.

Preindividuation Genetic Interventions in Relation to Identity

Recall Kripke's plausible thesis that none of us could have derived from a different sperm and egg than those from which he did in fact derive. Deriving from those precise gametes is necessary for the existence of the same individual in any possible world (that is, in the world as it actually is or in any possible counterfactual situation). But why? A plausible answer will suggest that preindividuation genetic interventions are likely to be numerical-identity-affecting. (Hereafter I will use *identity-affecting* as elliptical for *numerical-identity-affecting*.)

Consider two competing accounts. What we may call the *precise-original-genome account* requires exactly the same genome, the very same genes, for a given human individual to have originated. On this view, at least part of the reason that deriving from the same sperm and egg is so important is that different gametes would have entailed a different genome (and therefore different genetic information). Even if one can and does survive changes to one's genome *once one exists* – we survive mutations all the time – on the present view one could not have *originated* with a different genome. In a counterfactual situation, someone who originated from a different genome would have been someone other than you.

A second, more plausible account that elaborates on Kripke's thesis – call it the *imprecise-original-genome account* – draws support from the following thought experiment. Suppose the very same sperm and egg from which you derived had joined, leading to conception, but with one difference. Right before conception, a mutation in one of the gametes slightly changed the genome of that gamete and therefore of the later zygote. Alternatively, imagine that the mutation occurred after conception but before unique individuation. Suppose also, in either version of this scenario, that the mutation made no difference to your later phenotype. Intuitively, it seems that, despite this slight change of genome, you would have come to exist in this scenario. This claim remains fairly plausible even if we change the details so that a few preindividuation mutations occur, so long as they are insignificant, resulting in very little or no change to the phenotype. Assuming you would have come into being despite the slight genetic difference, your identity does not depend on deriving from a zygote with *precisely* the same genome; it depends on deriving from the same sperm and egg and *virtually* the same genome.

This raises a question. The present view holds that deriving from the same sperm and egg is necessary, but precisely the same genome is not. So would deriving from the same sperm and egg suffice for one to come to exist in any possible world *so long as most of the same material – the actual physical stuff (e.g., cytoplasm) – is present?* No, because surely you would not have come into being if the entire nucleus of the egg had been removed and replaced with another one. The original material of the sperm and egg plays some role in the identity of those particular sperm and egg, and therefore in your identity (because the numerical identity of those sperm and egg plays a role). Meanwhile, having virtually the same genome is also essential. How to specify further the necessary and jointly sufficient conditions for one of us to originate is an extremely complicated issue that we cannot pursue here. It suffices for our purposes to embrace the imprecise-original-genome account.[20]

What are the implications for preindividuation genetic interventions? If the precise-original-genome account were true, then any preindividuation intervention that changed genetic information would affect which organism comes into being. By contrast, the imprecise-original-genome account suggests somewhat less sensitivity on the part of the gametes and subsequent zygote. Some genetic changes prior to unique individuation would be identity-preserving, permitting the same individual to come into being as would otherwise have originated. On the other hand, the purpose of genetic therapy or enhancement is to cause a change in genotype that makes a *significant* difference to the later phenotype. This suggests that, in any successful genetic intervention, the genetic change matters. I will therefore assume that preindividuation genetic interventions that achieve their purpose are identity-affecting, resulting in the origination of a distinct human organism than would otherwise have originated.

---

[20] This discussion has focused on *necessary* conditions for one's existence in any possible world. Clearly, the conditions discussed here are not sufficient. For early in gestation, one has yet to be uniquely individuated. This was explained in the previous section with reference to the possibilities of monozygotic twinning and fusion. In addition to requiring derivation from the same zygote, which in turn requires derivation from the same sperm and egg and virtually the same genome, one's existence necessarily involves unique individuation as *some particular* human organism of the type that we are; one is essentially that human organism. But the criteria determining what counts as some particular human organism may be impossible to articulate fully, beyond ostension – literally, pointing – or rigid designation with a name (see Kripke, *Naming and Necessity,* Lectures I and II). I believe the same idea can be found in the medieval philosopher Duns Scotus's idea of *haecceity* or "thisness," the fundamental particularity of an existing thing.

Postindividuation Genetic Interventions in Relation to Identity

Consider now genetic interventions on uniquely individuated human organisms, for which I will conveniently reserve the term *fetuses*. Each of us originated as a fetus. How sensitive is a fetus's numerical identity? I submit that it is not very sensitive at all, that a given human fetus can survive many changes of genotype – including those likely to be attempted in genetic interventions for therapeutic or enhancement purposes. Call this the *robustness thesis*.

Each of us is essentially a particular living human organism. But nothing in the idea of a particular living human organism seems incompatible with genetic – or, for that matter, environmental – changes that alter the disposition to certain diseases, baldness, eye color, height, athletic ability, level of intelligence, or the like. Once one exists, one can change a great deal without going out of existence. When I was a child a neighbor was hit by a car, causing him to become mildly mentally retarded. Although the accident had a significant impact on his intellectual abilities and on the life he would lead, it surely didn't kill him, or otherwise end his existence, and replace him with a mentally retarded successor. Similarly, once someone comes into existence as a fetus, that individual can change substantially – whether due to genetic interventions or factors in the uterine environment (such as trauma to the pregnant woman's body) – without going out of existence. Once we exist, our numerical identity is quite robust through change, as suggested by the thesis that human persons don't go out of existence until biological death occurs (as argued in Chapters 2 and 4). Generally, when we change a lot, *we* change a lot; we become different people in a qualitative sense – with possible implications for narrative identity – but not in the sense of numerical identity. Failure to grasp this relatively simple point has, as we will see, vitiated much argumentation in the literature on prenatal genetic interventions.[21]

[21] Does the robustness thesis assert that no conceivable change to a fetus's genome would be identity-affecting? I intend nothing so radical. Imagine a genetic intervention that changed a human fetus's genome so greatly that a living creature in roughly the form of a fish resulted, before being transferred to an artificial womb and permitted to gestate. Could one of us survive transformation into such a creature? Intuitively, it may seem that we could not. But part of the difficulty of imagining that we would survive such transformation may stem from the difficulty of imagining the transformation itself. Rather than consulting our intuitions, we might consult our account of human identity, which states that each of us is essentially a particular human organism. With this point of departure, we ask whether a particular change is compatible with the continuing existence of the same human organism. Now one might say that something

The robustness thesis, as I intend it, suggests that (postindividuation) genetic interventions likely to be attempted in the name of therapy or enhancement will not be identity-affecting. Once we come into existence, our identity is sufficiently stable that foreseeable genetic interventions will not create numerically distinct human organisms. As noted earlier, quite a few commentators think otherwise. Why? Consider some samples.

Walter Glannon:

[G]ene therapy designed to correct or treat a cognitive or affective disorder would be more likely [than therapy not designed to treat a mental disorder] to alter one's identity. The manipulation of the relevant neurotransmitters or regions of the brain that generate and support mental life would directly affect the very nature of the mental states definitive of personhood and personal identity....[22]

Glannon's thesis that our numerical identity is sensitive to prenatal genetic interventions that have a significant impact on mental life is traceable to his implausible assumption of person essentialism, the thesis that we are essentially persons.[23] If we were essentially persons, then we would

in the form of a fish is obviously not a human organism, suggesting that such transformation would be identity-affecting. But matters are not so simple for two reasons.

First, it is debatable how to construe the species membership of the fishlike thing. Is a fish any living organism with a genome within the range associated with what are uncontroversially fish? Or is a fish any living creature reproduced from fish parents? We may similarly wonder whether the fishlike thing is human. Perhaps a human is anything that was reproduced from human parents and has not suffered biological death. This conception is radically flexible and would allow a human to persist through any phenotypic changes that are not lethal (in the usual sense). Another possibility, of course, is that a creature is not human unless its current genotype falls within some range associated with *Homo sapiens* – or perhaps, more broadly, the full range of past and present hominids. On this conception, the radical genetic intervention puts the human organism out of existence, replacing it with another type of creature.

A second uncertainty arises in this way. I argued in Chapter 2 that we are essentially human animals. But I clarified that the thesis was not intended to be precise about the boundaries of our kind. Maybe a human animal could survive transformation into another kind of animal. Perhaps we are essentially animals, or something more specific such as primates, but not essentially *human* animals (despite my adopting this phrase). The main thesis was that the criteria of our identity are biological, not psychological – a thesis that leaves somewhat open the boundaries of our essential kind. Thus, even if we could say with certainty that a radical genetic intervention would result in a creature that was not human, the biological view cannot automatically infer that such a transformation would be identity-affecting. Perhaps one of us lives so long as he hasn't died in the usual biological sense. I leave these issues open.

[22] *Genes and Future People* (Boulder, CO: Westview, 2001), pp. 81–2.
[23] Ibid., ch. 1. He clearly has in mind numerical, not narrative, identity because the former is the only kind of identity he addresses.

not come into existence until the human organism constituted or otherwise produced a person; we would *originate* as persons, "most plausibly in infancy,"[24] so (according to this reasoning) any significant genetic changes to the prepersonal organism would determine that a numerically distinct person came into being.

After carefully distinguishing the numerical identity of the organism and that of the person, Robert Elliot argues that prenatal genetic therapy (along with environmental factors) determines which person subsequently comes into being.[25] Somewhat similarly, Christopher Belshaw writes, "Gene therapy, performed on a human embryo, can affect the [numerical] identity of the person to be born."[26] But, unless we were essentially persons, what difference would it make which person comes into existence (whatever exactly that may mean)? If we are essentially human organisms, as I have argued, then genetic interventions could improve the quality of life of the affected individual once she becomes sentient.

Moreover, it's far from clear that changes in the prepersonal human organism would affect the numerical identity of the later person – even if we were essentially persons. For how should we understand the relationship between organismic identity and personal identity in this case?[27] Perhaps Glannon, Elliot, and Belshaw are assuming the *type-cast theory* of this relationship: *The person's identity is determined partly by the identity of the fetal organism and partly by some of the person's most distinctive properties.* Different persons could originate from the same fetal organism, in different possible worlds, depending on what happens to that organism and, in particular, what distinctive properties the person comes to have. On this view, if Plato's mother had dropped the infant who in fact grew into the man we know as Plato, so that the infant was brain-damaged and never took up philosophy years later, then the man we know as Plato would never have existed – because the brain-damaged man would have lacked too many of Plato's distinctive features. Consideration of counterfactual situations such as this one render the type-cast theory highly implausible. After all, the story just told is not only perfectly intelligible; it seems to be about *that same individual, Plato,* even if the truth of this story would

---

[24] Ibid., p. 26.

[25] "Identity and the Ethics of Gene Therapy," *Bioethics* 7 (1993): 27–40.

[26] "Identity and Disability," *Journal of Applied Philosophy* 17 (2000), p. 263.

[27] I got the idea of distinguishing the two theories described in this paragraph from a discussion in Ingmar Persson, "Genetic Therapy, Identity, and the Person-Regarding Reasons," *Bioethics* 9 (1995), p. 22.

mean that few people ever heard of Plato, and the history of philosophy turned out very differently from the way it actually turned out.[28] Much more plausible, I suggest, is the *origination theory: The person's identity is determined entirely by the identity of the fetal organism.* Rather than being essentially the author of *The Republic* or the greatest influence on Aristotle, Plato, that person, was whatever person (if any) "emerged" from the human organism that was, in fact, associated with the Plato we know. But the issue dividing these two theories of the organismic identity–personal identity relationship is motivated by the assumption that a person's identity differs from that of the associated organism, an assumption based on person essentialism.

Supporting the thesis of fragile prenatal identity in a different way, Noam Zohar concludes about fetuses that "persistence of genotype must generally be deemed a necessary condition for maintaining personal identity. Therefore, many proposals for [prenatal genetic interventions] should be excluded from the notion of therapeutic intervention."[29] According to Zohar, since prenatal genetic interventions designed to be therapeutic will usually aim to remove significant deficiencies in bodily or brain functions, such interventions will, if successful, result in the creation of an individual distinct from the one who was supposed to be helped – amounting to eugenics rather than therapy.[30] But Zohar's conclusion that "therapeutic" interventions would alter identity depends on his thesis that our identity, like that of a ship, is a matter of the functional organization of the entity in question.[31] In my view, by contrast, while the identity of a ship may concern, among other things, its functional organization, the identity of a human organism or any other living thing is a matter of the continuation of one and the same life.

Before turning to the ethics of prenatal genetic interventions, consider another possible basis for believing that prenatal genetic identity is fragile (even after unique individuation), as conveyed in this passage: "[T]he disease had a profound effect on the boy's identity. [T]he immune deficiency caused by some genetic flaw severely limited activities, experiences, and interactions of which he could partake and in which

---

[28] In this way, Kripke overturns the Frege–Russell descriptive theory of proper names (*Naming and Necessity*, pp. 27–31, 53–4, 58–60).

[29] "Prospects for 'Genetic Therapy' – Can a Person Benefit from Being Altered?" *Bioethics* 5 (1991), p. 275. He uses the term *person* broadly so that it applies to prenatal human organisms.

[30] Ibid., p. 288.

[31] Ibid., p. 283.

he could participate. [So it was] surely an important factor in making him who he was."[32] From this one might conclude that genetic "therapy" could put its intended recipient out of existence. But such a conclusion would probably rest on a conflation of numerical and narrative identity. To be sure, whether or not I have some terrible medical condition will affect who I am insofar as it affects my self-narrative. But that is considerably more plausible than, and does not entail, the claim that a genetic intervention that prevents such a condition affects numerical identity. It does not, on my view, because once a human organism has been uniquely individuated, its identity is robust.

ON THE ETHICS OF PRENATAL GENETIC INTERVENTIONS

So far we have considered (1) when in prenatal development one of us originates and (2) the relationship between prenatal changes – in particular, genetic interventions – and numerical identity. I have argued, first, that we originate at the time of unique individuation; and, second, that preindividuation genetic interventions are likely to be identity-affecting, whereas postindividuation genetic interventions will not. With this conceptual background, we turn to the ethics of prenatal genetic interventions.

Although prenatal genetic therapy (PGT) and prenatal genetic enhancement (PGE) have yet to be undertaken in human beings, they have been subject to much discussion. The broader public has expressed a mixture of prospective support for such genetic interventions, especially PGT, and objections to their development. The objections range from concerns about risks, to the charge that interventions on the genome are inherently immoral for "playing God," to the objection that PGT will express and reinforce a mainstream devaluation of persons with disabilities. In this discussion I will assume that PGT and PGE are not inherently immoral.[33] Moreover, I understand the important concerns

---

[32] Jeffrey Kahn, "Genetic Harm: Bitten by the Body That Keeps You?" *Bioethics* 5 (1991), p. 299. Kahn does not endorse the reasoning he articulates in this passage.

[33] The thesis that they are inherently immoral for "playing God" presumably rests on either (1) the thesis that God forbids these interventions, (2) the thesis that they violate some inviolable core of the human individual, or (3) the thesis that, by affecting numerical identity, they effectively decide who comes into being. My reply to (1) is that irreducibly religious assumptions are illegitimate bases for public policy. Chapter 6's reply to the charge that some enhancements would violate an inviolable core in the postnatal context may be extended to reply to (2) in the prenatal context. This section includes my reply to (3).

raised by disability advocates to be contingent on the social circumstances in which PGT and PGE would occur; these concerns, then, are compatible with my assumption that the interventions are not *inherently* immoral.[34]

Let us now consider a perspective – call it the *mainstream view* – that seems likely to be widely held among those who have considered the prospect of PGT and PGE and do not judge them inherently immoral:

There will probably be cases in which PGT is the only promising way to help someone, by preventing, eliminating, or ameliorating the effects of some disease or impairment. In such cases, if there is sufficient reason to believe that the promise of benefit to the individual on which PGT would be performed compensates for any risks associated with the therapy, PGT will be morally acceptable. By contrast, PGE does not prevent, eliminate, or ameliorate the effects of a disease or impairment. It aims merely at enhancing in the later person certain traits that would otherwise fall within the normal range. The desirability of such enhancement is almost certainly insufficient to justify the risks attending PGE. These risks include not only medical risks to the individual in question, but also social risks to him or her (e.g., a later feeling of having been commodified) and social risks to society at large (e.g., greater pressure on parents to choose PGE in order to keep pace with others, exacerbation of existing social inequities).

As suggested in the previous section, considerations of identity quickly complicate matters. Do these considerations, or other concerns, undermine the mainstream view?

Prenatal Genetic Therapy

The concept of therapy implies that the recipient of the intervention stands to benefit. But would prenatal genetic "therapy" really be therapeutic to the individual on whom (or which) it is performed? Our present concern is philosophical, not technological, so let's assume that the intervention would have its intended effect of preventing, eliminating, or ameliorating some condition. Now the salient issue is whether PGT would be identity-affecting. (I will use *therapy* to refer to the sorts of interventions that are ordinarily taken to be therapy, despite the identity-based doubts about their therapeutic status.)

---

[34] For samples of literature expressing such concerns, see Lynn Gillam, "Prenatal Screening and Discrimination against the Disabled," *Journal of Medical Ethics* 25 (1999): 163–71; and Adrienne Asch, "Prenatal Diagnosis and Selective Abortion: A Challenge to Practice and Policy," *American Journal of Public Health* 89 (1999): 1649–57. For an excellent discussion of these concerns, see Allen Buchanan et al., *From Chance to Choice: Genetics and Justice* (Cambridge: Cambridge University Press, 2000), ch. 7.

If PGT is not identity-affecting, then it benefits its intended recipient, qualifying as genuinely therapeutic. Moreover, there would be what some call a *person-regarding reason* for delivering PGT. That is, the moral basis for PGT would be therapeutic beneficence – a time-honored medical priority – which is directed to a particular identifiable individual standing in the role of a patient.[35] But, because the term *person-regarding reason* is potentially confusing in half-suggesting that the individual in question must already be a person when no such requirement is justified, I will substitute the less loaded term *individual-regarding reason*.

As noted in the previous section, some commentators believe that PGT is identity-affecting.[36] On this view, PGT does not benefit the individual on which it is performed because, by affecting identity, it eliminates this individual and creates a new individual in the former's spatiotemporal wake.[37] For this reason, the traditional individual-regarding reason for medical interventions, therapeutic beneficence, does not apply. Indeed, as there is apparently no individual-regarding reason for PGT, it proves to be far less important than most medical interventions – making it an unlikely candidate for public funding.[38] For

[35] Depending on one's view of identity and moral status, one may take the patient to be either the current prenatal individual, the later sentient fetus-cum-newborn, the later newborn, or the later person; if one takes one of these later beings or stages to be the patient, the current prenatal individual is treated *as if* it were the patient because current interventions will predictably affect the later patient.

[36] See, e.g., Zohar, "Prospects for 'Genetic Therapy'." Kahn argues that, even if Zohar is right about changes in prenatal identity, the best account of harm implies that certain genes are inherently harmful, vindicating roughly the usual way of distinguishing therapy and enhancement ("Genetic Harm"). Although his discussion of harm is very valuable, Kahn does not persuade me that his account overcomes Zohar's identity-based challenge.

[37] On this view, PGT would be beneficial only in the relatively rare case in which the prenatal individual is better off being eliminated than being permitted to survive due to extremely burdensome medical problems that continued existence would entail.

[38] After arguing that prenatal identity is relatively robust on the strength of the origination theory (see the preceding section), Persson contends that, even if this is mistaken and prenatal identity is fragile, there is nevertheless an individual-regarding reason favoring prenatal genetic interventions in identity-affecting cases: By bringing a new individual into existence, these interventions *benefit* that individual ("Genetic Therapy, Identity, and the Person-Regarding Reasons"). This thesis relies on the controversial premise that one can benefit by being brought into existence. Although I cannot fully defend my position here, I hold to the contrary that, while one can benefit by continuing to exist, coming into existence cannot benefit an individual because there is no determinate being who could either exist or not exist – no one who stands to benefit. Even if I am wrong here, my reply to the claim that PGT is not therapeutic because it is identity-affecting is independent of the questionable thesis that coming into existence is a benefit.

those who regard the current prenatal individual as having significant moral status, this conclusion is reinforced by an individual-regarding reason *not* to perform PGT: It eliminates someone with moral status.

In my view, as explained in the previous section, whether PGT is identity-affecting depends on the developmental stage at which it occurs. Prior to unique individuation – which, again, takes place at some point between a few days and about two weeks after conception – PGT is likely to be identity-affecting, whether performed on preconception gametes or on the zygote.

Nevertheless, this does not cast significant moral doubt on the procedure. For, at this stage, PGT does not eliminate a human organism of our kind. Rather, it prevents such a being from coming into existence – just as contraception does. Since there is no significant moral objection to preventing a human individual from coming into existence, as opposed to eliminating an existing one, the fact that preindividuation therapy affects identity does not count morally against it. Such therapy innocuously determines that a different human organism will originate than otherwise would have originated, rather than harming or otherwise affecting an existing one.

Meanwhile, there is a substantial moral reason favoring preindividuation genetic therapy: If successful, it will prevent whoever eventually comes into being from suffering from the targeted genetic disease or impairment. Because the therapy is identity-affecting, this is not an individual-regarding reason in the sense of being directed toward an already existing individual. Nevertheless, that it prevents significant suffering without harming anyone, or violating anyone's rights, counts heavily in its favor. Assuming that other sorts of concerns – such as those voiced by disability advocates – do not justify a condemnation of PGT across the board, and that concerns about safety are responsibly taken into account, it would appear that preindividuation genetic therapy is morally acceptable, praiseworthy, and in some circumstances perhaps even mandatory. If reasonably promising from a technical standpoint, it has a strong claim on public funding.

After a human organism has been uniquely individuated, PGT is not identity-affecting. One and the same life will continue as the individual's genome undergoes a relatively modest change. So the worry that PGT (even after unique individuation) is not really therapy, but rather a form of eugenics – a worry based on the assumption that PGT replaces the original human organism with a numerically distinct, "better" one – proves unfounded. The mainstream understanding that PGT benefits the already existing human organism is correct in postindividuation cases. In these cases, then, considerations of identity in no way vitiate the mainstream view on PGT. Other factors such as feasibility and safety being equal, it deserves the same approval and priority that other medical therapies enjoy.

Prenatal Genetic Enhancement

The mainstream view judges PGE to be much less defensible than PGT and probably, all things considered, unjustified. We explored enhancements at length in Chapter 6, but not in the prenatal context. PGE seeks to change gametes, the zygote, or the uniquely individuated human organism with the aim of endowing the later person with some desired trait such as increased height, a different eye color, superior athletic ability, or a more formidable memory. PGE differs from PGT in not being directed at a genetic disease or impairment.

Since our identity-related concern is philosophical, assume for the moment that a particular instance of PGE is likely to be feasible – safe and effective in bringing about the desired genetic (and later phenotypic) change. Now, whether PGE would benefit its recipient depends, as with PGT, on whether the intervention would be identity-affecting. And that, once again, depends on when it is performed.

If PGE targets the preconception gametes or the zygote prior to unique individuation, its alteration of genetic information is likely to determine which human individual will later originate. But because there is not yet a uniquely individuated human organism – a being of our kind – it does not put such a being out of existence. Moreover, it has the advantage of ensuring that whichever human individual comes into existence has some desirable characteristic. While this is not an individual-regarding reason for PGE, as there is no already existing individual who will benefit, it does eventually bring about an advantageous trait without causing harm or violating anyone's rights, a point that speaks at least somewhat in its favor. There is therefore no cogent identity-based moral argument against preindividuation genetic enhancement.

After a human organism has been uniquely individuated, PGE is not identity-affecting. One and the same individual will undergo a change in genetic information, with predictable effects on the later phenotype.

So postindividuation genetic enhancement faces no intelligible moral objection on grounds of identity.

But, of course, some moral concerns about enhancement technologies have nothing to do with identity. We explored such concerns at length in Chapter 6. While we found that they did not justify a blanket prohibition of the use of enhancement technologies, or some broad category of them, the cumulative doubts about genetic enhancement were more troubling than those regarding cosmetic surgery and cosmetic psychopharmacology, the two other broad categories of enhancement technologies we examined in detail. Concerns about risks and about possible injustice were prominent.

Concerns about risks are amplified in the prenatal context because we know far less about the technology of prenatal genetic interventions than the relatively little we know about postnatal genetic interventions. Further, as with the latter, in the prenatal context the risk–benefit ratio of attempted genetic enhancements is likely, in most cases, to be worse than the risk–benefit ratio posed by attempted genetic therapy. Once again, therapy responds to a highly undesirable status quo, some disease or impairment or disposition thereto, so the benefits of successful therapy are very considerable, quite possibly outweighing the risks, especially if nongenetic means of treatment are lacking. But enhancement typically responds to a much more tolerable status quo – some suboptimal functioning within a normal range – so the benefits of successful enhancement will usually be smaller than those of therapy, with no expectation of a proportionate reduction in risk. Moreover, the social risks unique to enhancement, as discussed in Chapter 6, would be present. Further, since a market in PGE will make it more readily, or exclusively, available to the wealthy, such a market will probably exacerbate existing social inequities and injustice. Finally, there is no possibility of obtaining the consent of the individual to be affected by PGE. (While this is also true with PGT, the latter will often be in the affected individual's best interests if nongenetic treatments are lacking.) By contrast, competent adults contemplating genetic enhancements for themselves can provide voluntary, informed consent. On this score, then, PGE is far more problematic than genetic enhancement involving competent adults and at least some instances of PGT.

Nevertheless, there is reason to think that *some* forms of PGE could eventually pass moral muster. Imagine that a type of PGT, whose target is a specific cognitive deficiency, has proved safe and highly effective. Meanwhile, a related form of genetic *enhancement* – improving an aspect of cognitive performance in those who qualify as cognitively normal – has proved safe and effective in competent adults. Soon, an intervention that is ambiguous between therapy and enhancement (but is promoted as therapy) proves safe and effective, and then becomes widely used, among teenagers and even children. It is, in fact, the same intervention as what was classified as enhancement, though it is used in a different population: those who show cognitive deficits but arguably lie within the normal range. Let us even assume that the benefits in this group and among those who are clearly within the normal range are very extensive, clearly greater than the benefits of some previously accepted genetic therapies. Now scientists propose to study what is essentially the same intervention, although it would be administered prenatally, to fetuses displaying no signs of a predisposition to cognitive deficits. This would be an instance of PGE. It would be very similar to both the PGT and the postnatal enhancement that have already proved safe and effective. As far as I can see, there are no compelling arguments for ruling it morally out of bounds.

One might reply, however, that concerns about exacerbating injustice recommend opposing even this form of PGE. But the force of this legitimate concern is limited in at least two ways. First, the scenario just described could take place in a society that has universal access to health care and otherwise meets requirements of justice in distributing wealth, the means of subsistence, and access to education. (I believe this is true, or nearly true, in Canada, most of Western Europe, Japan, and various other countries, but not the United States.) The inequalities resulting from residual free markets in health care, education, and other goods in this country seem tolerable from the standpoint of justice, so justice-based concerns in this country would not condemn the imagined PGE. Second, even in countries guilty of gross distributive injustice, the *use* of this enhancement might not be wrong or blameworthy. Against this background of gross injustice, it might be wrong for a government to provide public funding for developing this technology; or, in the absence of public funding, perhaps a responsible society would find ways to discourage private funding for this technology until the national priority of universal access to health care (decent education, etc.) is achieved. But suppose that, despite compelling justice-based objections, the technology emerges in the nation in question, whether through public or private funding. Now parents face a choice of whether to avail themselves of it for their children-to-be. Once this choice is available to them, individuals may choose PGE for their offspring – and, as far as I can see, this choice *may* be a permissible and decent one.

These reflections therefore recommend openness, in principle, to some forms of PGE. This openness, however, is consistent with a strong *presumption* against PGE, especially in countries featuring gross distributive injustice, and with the claim that PGE can make no very strong claim on public funding. I support both this presumption and the latter claim. If my position is correct, then the mainstream view about PGE is right to assert a presumption against PGE but wrong to preclude it. But I venture no prediction as to whether any type of PGE will, in fact, ever emerge while appropriate moral requirements are satisfied.

Having addressed prenatal genetic interventions with special attention to the human organism's origins and identity, in the final two sections we will investigate certain types of reproductive choices with the same focus on our origins and prenatal identity.

REPRODUCTIVE CHOICES THAT MAY AFFECT IDENTITY: THE NONIDENTITY PROBLEM

PGT and PGE are future possibilities. Current genetic technologies permit reproductive choices based on genetic testing, the prenatal diagnosis of genetic disposition to certain diseases and impairments. As our understanding of the human genome advances, our ability to diagnose genetic dispositions to these conditions will also grow. Other (nongenetic) medical technologies can also inform a woman about the medical advisability of her becoming pregnant or continuing a pregnancy.

Prospective parents may elect not to attempt pregnancy if genetic or other medical information indicates serious risks to a fetus. Alternatively, they may choose to terminate a pregnancy on learning that their fetus carries a gene for cystic fibrosis, spina bifida, Huntington's disease, Down syndrome, or another disease or impairment. Nevertheless, most of these conditions are compatible with lives that are quite worthwhile from the standpoint of the subjects living them. For example, a person carrying the gene that causes Huntington's disease might enjoy forty or more years of normal health before developing symptoms; a person with Down syndrome might lead a very satisfying and full life that includes friendships, employment, and long-term partnership despite encountering certain challenges caused by mild mental retardation.

By contrast, some genetically transmitted diseases or impairments may predictably cause so much suffering and impairment to afflicted individuals that the conditions preclude a life worth living.[39] In these cases, it would be cruel knowingly to bring someone with the condition into existence. The charge of *wrongful life* – that the parents or physicians who negligently permitted such an individual to be born *thereby harmed, or at least wronged*, that individual – is meaningful only where the predicted quality, and perhaps length, of life are so poor that existence in this condition will clearly be a net burden, and not worthwhile, to the affected individual. I accept the judgment that some rare conditions are incompatible with a life worth living. To be sure, the coherence of the charge of wrongful life has been challenged on the basis of the conceptual difficulties of comparing the affected individual's life of suffering with any possible condition that represents a preferable alternative. Nonexistence – as entailed by death or by never coming into existence – is not a condition. Although I believe claims of wrongful life are coherent, I will not pursue this difficult issue here.[40] The topic of this section is conditions that permit lives worth living, although they are inherently disadvantageous in comparison with species-typical functioning or normal health.[41]

As Derek Parfit first noted in a discussion that has generated copious commentary, causing individuals to exist in such conditions provokes a "non-identity problem."[42] Rather than offer a comprehensive review of this literature, I will explain the problem and address it in a way consistent with the theoretical framework developed in this book, noting in passing some other attempts to resolve the problem.[43] We may borrow from Dan Brock a revised version of the vignette Parfit used to introduce the topic and two other vignettes for purposes of comparison – along

[39] Possible examples include Lesch-Nyhan syndrome, Tay Sachs disease, and dystrophic epidermolysis bullosa.

[40] For good discussions, see Buchanan et al., *From Chance to Choice*, pp. 232–42; and Jeff McMahan, "Wrongful Life: Paradoxes in the Morality of Causing People to Exist," in Jules Coleman and Christopher Morris (eds.), *Rational Commitment and Social Justice* (Cambridge: Cambridge University Press, 1998), pp. 215–16.

[41] If we expand the relevant sense of *conditions* to include socioeconomic circumstances, we may also consider cases in which individuals are brought into existence in conditions that fall short of what is required for responsible parenting, as when an immature teenager lacking social supports has a child. Our discussion focuses on medical conditions.

[42] *Reasons and Persons* (Oxford: Clarendon, 1984), ch. 16.

[43] For a good review, see David Wasserman, "Personal Identity and the Moral Appraisal of Prenatal Therapy," in Lisa Parker and Rachel Ankeny (eds.), *Mutating Concepts, Evolving Disciplines* (Dordrecht, the Netherlands: Kluwer), pp. 248–58.

with the term *wrongful handicap cases* to designate such cases, in which the handicap, though significantly disadvantaging in comparison with normal functioning or health, is compatible with a life worth living.[44]

> *C1 (preconception):* A doctor informs his patient and her husband[45] that they should delay attempts to conceive because the woman has a medical condition that would predictably cause any child she bears to be mildly mentally retarded. If she takes a safe medication for a month, she can later get pregnant and give birth to a healthy child. Because she and her husband do not follow this advice, they achieve pregnancy in the next few weeks, leading to the birth of a mildly retarded child.
>
> *C2 (prenatal):* A doctor informs his pregnant patient and her husband that she has a medical condition that will cause mild mental retardation in the child-to-be unless she takes a safe medicine. Because the patient does not take the medicine, she later gives birth to a mildly retarded child.
>
> *C3 (neonatal):* A doctor informs two parents that their newborn child has a condition that will predictably lead to mild mental retardation unless they give him a safe medicine. Because the parents do not administer the medicine, the condition causes mild mental retardation in their child.

Common intuitive reactions to C1 include the claims that (1) the parents act wrongly and, more specifically, (2) they *wrong their child* by not preventing his handicap when they could easily have done so. Most people are also likely to believe that, morally, C2 and C3 are relevantly similar to C1. Parfit calls this judgment of relevant moral similarity the *no-difference view*. But considerations of numerical identity show that C1 is importantly different from C3 in a way that makes it difficult to justify certain judgments of common morality, including the claim that the child in C1 is wronged. Later it will become clear that considerations of identity cast different lights on C2, depending on the stage of pregnancy in which a decision about medicine is made.

The parents' failure to prevent their child's handicap in C3 is a paradigmatic case of child neglect. Had the parents provided their newborn baby

---

44 "The Non-Identity Problem and Genetic Harms – the Case of Wrongful Handicaps," *Bioethics* 9 (1995), p. 270.

45 These cases consistently present the mother as the responsible party. Presumably, the father has some role in the pregnancy, so I present the cases in a way that makes the father's responsibility explicit.

a safe medicine, as they were advised, he would not have become mildly retarded. Reflecting on his parents' earlier choice when he is older, the child (adolescent, adult) could justifiably claim that his parents wronged him with their neglect. But C_1 is different. For on any leading theory of numerical identity, had the parents delayed pregnancy for more than a month, a different sperm and egg would have united in fertilization. Because of the necessity of origins – the fact that none of us could have derived from a different set of gametes than those from which he or she did derive – the result of the parents' responsible choice would have been the birth of *someone other than* the individual who was in fact born in C_1.[46] The mildly retarded individual in C_1 is therefore in no position to complain about his parents' irresponsibility, for he would never have existed had they acted differently. We need not decide whether coming to exist is a benefit in order to accept this point. For, surely, coming to exist with a life that is worthwhile (albeit with a handicap) *cannot be worse* than never existing. If coming to exist with a handicap were worse than never existing, the life would not be worthwhile – but this is so only in wrongful-life, not wrongful-handicap, cases. It would appear, then, that the parents in C_1 act wrongly *but do not wrong their child.* How to make sense of this is the nonidentity problem.

For several reasons, the problem emits an air of paradox. First, as just noted, many people intuitively believe that the parents do wrong their child in C_1. No doubt some will abandon that belief on realizing the point about nonidentity: that the child born is not the one who would have been born had the parents acted responsibly. Still, a common considered judgment is that wrongdoing requires a victim, yet in this case of apparent wrongdoing there seems to be no victim. Second, the parents' behavior in C_3, in which the child is clearly wronged – since he would have existed even if the parents had acted responsibly – seems no worse than their conduct in C_1.[47] Now, if the child really is wronged in C_1, contrary to what

[46] Interestingly, the thesis of the necessity of origins is incompatible with one traditional theory of identity: that each of us is an immortal soul. For there is no reason to assume that a particular soul must "attach" to a particular set of gametes. Perhaps a soul could be "ready for delivery" to the next fertilization whenever it may occur. The plausibility of the thesis of the necessity of origins counts heavily against substance dualism (see Chapter 2) and helps to show why the immortal-soul view of human identity is not a leading contender. Cf. Kripke, *Naming and Necessity,* p. 155, n. 77.

[47] Here I assume that the child who becomes retarded existed as a newborn, so that it is not some successor to the newborn who bears the handicap. Nearly everyone – including philosophers steeped in identity theory – holds that each of us existed at birth. Hence the *newborn problem* for person essentialism, which implies that we did not exist until a person emerged (see Chapter 2).

considerations of identity seem to suggest, this would dissolve the second puzzle: The parents' actions in C1 and C3 would be equally wrong because the child would be (equally) wronged in both cases. But if, as Parfit, Brock, and I think, the child is not wronged in C1, there remains the puzzle of why the parents' behavior in C3 – which has a victim (someone harmed through neglect and thereby wronged) – seems no worse. Further, if the child isn't wronged in C1, how is the parents' behavior wrong? How can it be wrong to do something that wrongs no one?[48]

In the context of reproductive decision making, the nonidentity problem poses a threat to certain widely accepted moral judgments. We generally believe that it is wrong for prospective parents knowingly to bring someone into the world with a handicap when they could – with no substantial costs to themselves or anyone else – bring someone into the world without any handicap. Many people, of course, oppose abortion on moral grounds, so, for them, abortion is an unacceptable means of not bringing an individual with a handicap to term. So let us restrict our point to preconception cases such as C1. Nearly everyone, including opponents to abortion, would judge that the parents act wrongly in C1, where merely postponing efforts to conceive would be sufficient to avoid bringing someone into the world with a handicap. By raising doubts about our intuitive reaction to C1 and similar cases, the nonidentity problem threatens to undermine widely held attitudes about obligations to prevent handicaps.

But the importance of the nonidentity problem extends much further. After examining several additional cases (which we need not review), Parfit contends that the part of morality concerned with benefiting and harming – or, more precisely, with the production of better or worse states of affairs (since the concepts of benefit and harm arguably do not apply in nonidentity cases) – needs radical revision. He refers to this part of morality, rather broadly, as *beneficence*, not distinguishing beneficence from nonmaleficence, as is common today; let us adopt this convenient shorthand for present purposes. According to Parfit, we must understand the morality of beneficence not in *person-affecting terms* – that is, in terms of an action's effects on particular determinate individuals – but in the *impersonal* terms of how much good and evil an

---

[48] While one might suppose that the parents will be burdened by having to care for a handicapped child, inviting the paternalistic charge that they wrongly disadvantage themselves, we might reject this charge on grounds of respect for autonomy. Moreover, we could surely add details to C1 such that even the parents are no worse off for their decision – without negating the sense that they acted wrongly.

action brings into the world. (Because sentient nonhuman animals, who are not persons, can be benefited and harmed, and because we come into existence before we are persons, I will substitute *individual-affecting* for *person-affecting.*)

The theoretical options are clearer in light of Parfit's distinction between *same-individuals* (his term is *same-people*) choices and *same-number* choices.[49] In same-individuals choices, the same individuals will exist no matter which alternative action is taken. C3 is a same-individuals choice because the same child will exist whether or not the parents administer medicine as advised. In same-number choices, the same number of individuals will exist no matter which action is taken, but the identity of at least one individual who will exist is affected by the choice. C1 is a same-number choice because, although either alternative will result in one child, a different child will come into being, depending on which alternative is selected. Contrasted with both same-individuals and same-number choices are *different-number* choices, which affect the number of individuals who will exist. Although cases like C1–C3 do not involve different-number choices, other sorts of cases – ranging from decisions about whether to have children at all to policy choices that influence family planning – clearly do.

Consider now an implication of Parfit's proposal to view beneficence impersonally: While it matters that the parents in C1–C3 bring about states of affairs that are less good than other states of affairs they could easily have brought about, it does not matter further in C3, which involves a same-individuals choice, that *a particular individual* is thereby made worse off. That the result is significantly suboptimal matters, but the fact that someone is harmed carries no independent moral importance. That's counterintuitive. Now the thesis that we must construe beneficence in impersonal terms may be welcome, at a general level, to utilitarians, whose theory understands all of morality in terms of the production of good and bad states of affairs. But even utilitarians are accustomed to thinking of morality in individual-affecting terms – in terms of how particular individuals are made better or worse off, harmed or benefited. Furthermore, utilitarians will join others in chafing at some of the apparent implications of Parfit's proposal in population policy and other areas. Parfit himself calls one of these implications, to which he devotes a whole chapter, the *repugnant conclusion*: roughly, that for any population of people, each with a very high quality of life, there is some much greater

[49] *Reasons and Persons*, p. 356.

possible population whose existence would be morally preferable, even though the lives of its members would be barely worth living.[50] Although we will say little about the implications for ethical theory of Parfit's proposal and alternative solutions to the nonidentity problem, by now the reader should have some sense of its vast importance.

Back to our cases. Some commentators refuse to take seriously the implication of the necessity of origins that, had the parents delayed pregnancy in C_1, a different child would have been born – entailing that the child who actually was born was *not* harmed by being brought into existence with a handicap, since he could not have existed without it and has a worthwhile life.[51] Even less promisingly, one commentator has claimed that, in choices that determine whether a particular individual will come into existence – *genesis choices* – only the interests of actual (already existing) people matter; the interests of possible people like the retarded child in C_1 are morally irrelevant.[52] This implies that the only thing wrong with the parents' behavior in C_1 is creating the costs to themselves and society of having a mildly retarded child rather than a cognitively normal one. On this view, the fact that the handicap was avoidable by delaying pregnancy is unimportant. I find this implication highly implausible. Indeed, I think the only reason that the parents' behavior in C_1 is *clearly* wrong is the fact that they could have easily avoided the handicap by delaying pregnancy. So I suggest that we regard the behavior in C_1 as wrong for (at least) this reason while unflinchingly accepting that the child who is later born has no grounds whatsoever for complaint. He has not been harmed or in any other way wronged by being brought into existence in a handicapped condition. Because he has a worthwhile life, and never could have existed in a preferable condition, he is simply not a victim.

Before turning to considerations that might support a different conclusion, consider the ambiguous case C_2. Is the individual who comes to exist in this case the same individual who would have existed had the mother taken the medication while pregnant? According to the view defended here, we come into existence early in pregnancy, at the time of unique individuation (somewhere between a few days and two weeks after conception). Therefore, taking the medicine after this time would

---

[50] Ibid., ch. 17.

[51] See, e.g., Ronald Green, "Parental Autonomy and the Obligation Not to Harm One's Child Genetically," *Journal of Law, Medicine & Ethics* 25 (1997), p. 8.

[52] David Heyd, *Genethics* (Berkeley: University of California Press, 1992), ch. 4.

not be identity-affecting due to "robust" identity. But taking the medicine before this time presumably would be identity-affecting by modestly altering the nature of the human organism that later originates. So $C2$ raises the issue of nonidentity only very early in pregnancy. After this point, $C2$ is relevantly similar to $C3$ in that the parents' behavior will not affect identity.

Unlike $C2$, $C1$ features nonidentity straightforwardly and without qualification. Delaying pregnancy would have resulted in a different child. So, again, it appears that the mildly retarded child who comes into existence has no grounds for complaint about his handicap and has not been wronged by his parents. But this apparent implication, which some (not I) continue to find counterintuitive even after fully accepting the fact of nonidentity, can be challenged in subtle ways. Each challenge attempts to explain the wrongness of the parents' action in individual-affecting terms.

One challenge stresses the plausible claim that, in creating a life, we have obligations beyond creating a life worth living. We must make every reasonable effort to provide the individual with certain goods appropriate to the life of a human person, which go beyond what is necessary for a life to be (barely) worth living. Therefore, we can wrong people we create by not meeting our responsibilities to them – even in nonidentity cases.[53] I agree with the premise about our obligations. Certainly, children can have worthwhile lives despite their parents' doing little to foster healthy self-esteem, permitting excessive television watching and not encouraging more rewarding activities, and failing to save a dime for their college educations. But parents owe their children more than this. The sticking point of the argument concerns the words *to them*. For in nonidentity cases, such as $C1$, while the parents act irresponsibly, they do not fail to meet their responsibilities *to their actual child*. The child in $C1$ could not cogently say, "You failed to meet your responsibilities to me," because the relevant responsibility – delaying pregnancy so as to avoid mild retardation – is not one the parents could possibly have met *to him*.

A related challenge to the claim that the parents do not wrong their child is as follows. We should embrace the nonconsequentialist maxim that claims of having been wronged should rest not on what happened

---

[53] See F. M. Kamm, "Genes, Justice, and Obligations to Future People," *Social Philosophy and Policy* 19 (2002), pp. 373–9. Cf. Michael Bayles, "Harm to the Unconceived," *Philosophy and Public Affairs* 5 (1976): 292–304.

to a victim as a result of the agent's action, but on the character of that action. With this focus, rather than asking whether the child in C1 is worse off than he would have been had the parents delayed pregnancy, we should ask whether they acted responsibly. In cases like C1, parents fail to meet their responsibilities to their children, thereby wronging them.[54] But, once again, there is a problem with the referent of the alleged wrong. As the child in C1 has no intelligible grounds for complaint regarding his parents' treatment of him, there seems to be no coherent basis for saying that he was wronged. Moreover, although I will not develop this point, I am skeptical of any ethical view that is so indifferent about what happens in the world as a result of our actions. While the consequences of our actions are not all that matter in the ethical evaluation of what we do, they are a significant consideration.[55]

These challenges stress the distinction between harming someone and wronging her.[56] Although the distinction is valid – for example, one can wrong another by disrespecting her autonomy without harming her – it does not help in C1 and similar same-number cases, where the fact of nonidentity negates not just the claim that the individual brought into being has been harmed, but also the claim that she has been wronged. In these cases, the agents act wrongly without wronging anyone in particular.

But cannot the child in C1 (coherently) resent his parents for conceiving him in culpable indifference, even if he is glad they did because he appreciates his life?[57] Certainly the child can judge that the parents acted wrongly. If he resents them for this, in order to be intelligible his resentment must take a highly impersonal, moral form: resenting them for acting wrongly. His resentment cannot include a judgment of having been wronged.

Having found no cogent critique of the claim that the parents in C1 do not wrong their child, who is not a victim, I will assume that this claim is correct. It does not follow, however, that there is no moral difference among cases like C1–C3. One possibility is that the parents' actions in the same-individuals cases, C3 and C2-postindividuation, are *slightly* worse,

---

54 Rahul Kumar, "Who Can Be Wronged?" *Philosophy and Public Affairs* 31 (2003): 99–118.

55 Kumar recommends contractualism as a basis for this nonconsequentialist line of argument (ibid.). For a devastating critique of contractualism, see Martha Nussbaum, *Frontiers of Justice: Disability, Nationality, Species Membership* (Cambridge, MA: Harvard University Press, in press).

56 So does James Woodward in one of the earliest efforts to respond to the nonidentity problem without embracing impersonal principles of beneficence ("The Non-Identity Problem," *Ethics* 96 [1986]: 804–31).

57 David Wasserman raised this question in correspondence.

morally, than in the same-number cases, C1 and C2-preindividuation. The difference is that in the same-individuals cases there is a victim who has grounds for resentment about his parents' neglect and its impact on his life. His right to responsible care from his parents was violated. While the children in the same-number cases may believe the same about themselves, their belief is incoherent and, in principle, extinguishable by lucid reflection. So perhaps a *slight-difference view* is more plausible than the no-difference view.[58] If so, the fact that the difference is very slight would help to explain why, intuitively, the parents' actions in C1 and C3 strike us as equally problematic, even after we take into account the fact of nonidentity in C1: The difference is too slight to notice intuitively. (If, alternatively, there really is *no* moral difference, that would explain the lack of intuitive difference slightly better.) In any case, while we may acknowledge a slight difference, we could follow Parfit in accepting impersonal principles – at least for same-number cases like C1.

But the slight-difference view implies that we should not construe beneficence in impersonal terms for all types of cases. In same-individuals cases, we should take into account whether particular individuals have been made worse off than they otherwise would have been and whether their rights have been violated. On the present view, then, beneficence is subject to both impersonal and individual-affecting principles. What is the relationship between the two types of principles within an overarching view of beneficence or, more broadly, within ethical theory? This is an enormously complex topic, whose resolution is beyond the aims of this book, but a few points are worth making here.

First, in same-individuals cases, which represent most of the cases we encounter in daily life, our principles and rules pertaining to beneficence should take an individual-affecting form. In same-number cases, where the agent's choice affects who comes into existence, impersonal principles should govern. Brock has articulated a plausible impersonal principle:

Individuals are morally required (or have a moral reason) not to let any child or other dependent person for whose welfare they are responsible experience serious and inadequately compensated harm or loss of benefit, if they can act so that, without affecting the number of persons who will exist and without imposing

---

[58] Cf. Buchanan et al., *From Chance to Choice*, p. 254. The authors fail to note the issue of when we originate and, consequently, group cases like C2 (regardless of when during pregnancy the choice is made) with cases like C3. Also, I don't know whether they would accept the claim that the difference is slight. McMahan argues vigorously that, if there is a moral difference of the sort proposed here, it cannot be more than slight ("Wrongful Life," pp. 243–4).

substantial burdens or costs or loss of benefits on themselves or others, no child or other dependent person for whose welfare they are responsible will experience serious and inadequately compensated harm or loss of benefit.[59]

Both individual-affecting and impersonal principles count. In comparing cases like C1 and C3, impersonal considerations in the former and individual-affecting considerations in the latter have roughly equal moral weight; if there is a difference here, it is slight. In other contexts, individual-affecting considerations may have vastly more weight. For example, it is generally far worse to allow someone enjoying a good life to die than to fail to bring into existence someone who is likely to enjoy a good life.[60]

One of the most interesting features of the emerging theoretical picture is that, as McMahan has suggested, impersonal and individual-affecting considerations are not only distinct and mutually irreducible; they are *nonadditive*.[61] That is, when both apply, the individual-affecting considerations carry the full moral weight of the pair. That is why the parents' behavior in C3, which is individual-affecting but could also be assessed in terms of impersonal considerations (as we assess C1), isn't far worse than their behavior in C1. If both individual-affecting and impersonal considerations retained the importance they carry when each applies in the absence of the other, C3 would be *massively* worse than C1. But it is not. The individual-affecting principle kicks the impersonal principle off the scale when it is time to weigh in.

How can this theoretical picture be elaborated to account for different-number choices, in which different options would cause different numbers of individuals to exist? The impersonal principle stated earlier, designed for same-number cases, captures the basic idea that we should prevent significant harm and/or promote well-being when we can do so without significant costs. But extending this idea to different-number cases may lead, as Parfit worries, to the conclusion that we should increase total well-being – if necessary, by massively increasing the population, thereby greatly lowering average well-being, so long as individuals' lives are still worth living. In response to this problem, Buchanan et al. state that "[o]nly person-affecting principles seem likely to avoid [such

59 "Preventing Genetically Transmitted Disabilities While Respecting Persons with Disabilities," in David Wasserman, Jerome Bickenbach, and Robert Wachbroit (eds.), *Quality of Life and Human Difference* (Cambridge: Cambridge University Press, in press).
60 McMahan, "Wrongful Life," p. 244.
61 Ibid., pp. 243–4.

implications], since only they require that a reduction in suffering or an increase in happiness be to a distinct individual."[62] But this neglects the fact that public policies that will predictably affect the number of individuals who will come into existence will also probably affect the identities of many of those individuals. In such cases, either individual-affecting principles do not apply or they apply only to some of the individuals to be affected by the choice, apparently requiring appeal to impersonal principles to deal with those whose identities will be determined by the choice. Either way, contrary to Buchanan et al., individual-affecting principles hold little promise to guide us in these cases. How ethical theory should approach them remains mysterious.

ON ABORTION

No topic is more controversial than the morality of abortion, to which we now turn. Our discussion will focus on the question of whether *early abortions* are morally permissible. By *early abortions* I mean abortions that occur before the fetus has become sentient, acquiring the capacity for consciousness; by *late abortions* I mean abortions of sentient fetuses. Neurological evidence suggests that a fetus becomes sentient at some time between five and seven months' gestation.[63] But the *vast* majority of abortions – about 99 percent in the United States – occur before five months.[64] So, a justification of early abortion would, very importantly, demonstrate the permissibility of most of the abortions that girls and women actually seek.

One recent novel effort to justify early abortion is McMahan's argument from personal identity.[65] According to his embodied mind account, we are essentially minded beings or beings with the capacity for

---

[62] *From Chance to Choice*, p. 254.

[63] See, e.g., Mary Seller, "The Human Embryo: A Scientist's Point of View," *Bioethics* 7 (1992): 135–40; J. A. Burgess and S. A. Tawia, "When Did You First Begin to Feel It? – Locating the Beginning of Human Consciousness," *Bioethics* 10 (1996): 1–26; Julius Korein, "Ontogenesis of the Brain in the Human Organism: Definitions of Life and Death in the Human Being and Person," *Advances in Bioethics* 2 (1997), pp. 25–6; and Vivette Glover and Nicholas Fisk, "Fetal Pain: Implications for Research and Practice," *British Journal of Obstetrics and Gynaecology* 106 (1999): 881–6.

[64] According to the Alan Guttmacher Institute (www.agi-usa.org/pubs/fb_induced_abortion.html), only 1 percent of abortions were performed in the twenty-first week or later in the United States in 1998 (the most recent year for which the institute provided data).

[65] *The Ethics of Killing: Problems at the Margins of Life* (Oxford: Oxford University Press, 2002), pp. 267–9.

consciousness. Therefore we originated rather late in pregnancy, when neurological development generated this capacity. So, if the embodied mind account is correct, then early abortions do not kill beings with significant moral status, making these abortions "relevantly like contraception and wholly unlike the killing of a person."[66] Surely this is one of the most elegant defenses of abortion (or at least early abortions) ever conceived. But in Chapter 2 we found the embodied mind account to be an inadequate theory of our essence and identity, suggesting its inability to justify early abortion.

A persuasive defense of early abortion must not rest on a highly contestable theory of human identity. Moreover, it must be able to reply effectively to the strongest arguments *against* abortion. Of course, many antiabortion arguments are religious in nature. But a defense of early abortion need not respond to religious arguments on their own terms. For there are no religious theses that are so compelling that it would be unreasonable to deny their truth. That is, even if certain religious theses are true, it is not the case that each of us can be expected to embrace them on the basis of publicly available evidence. (That provides one justification for the right to freedom of worship.) But in order to be compelling, a moral argument must flow from assumptions that it would be unreasonable to deny. Thus, any persuasive argument against abortion must be persuasive in nonreligious terms.

Philosophers and bioethicists generally regard *the Future-Like-Ours Argument* (FLOA), developed in an article by Donald Marquis, as the strongest nonreligious argument against abortion.[67] An indication of the respect with which this argument is accorded is the frequent reprinting of Marquis's article in ethics anthologies – quite often as the only nonreligious representative of the antiabortion view.

After sketching the FLOA, I will suggest that its most plausible specification will incorporate the view of our origins defended in this chapter. Next, I will explain why what is often thought to be the most effective strategy for answering the FLOA – appealing to the Good Samaritan Argument – invites significant doubts about the latter's soundness.[68] But

---

[66] Ibid., p. 267.

[67] "Why Abortion Is Immoral," *Journal of Philosophy* 86 (1989): 183–202.

[68] Notably, Marquis assumes without argument that "the moral permissibility of abortion stands or falls on the moral status of the fetus" (ibid., p. 202). This is tantamount to assuming that the GSA is wrong in claiming that abortion is permissible *even if* fetuses have the same moral status as paradigm persons and, therefore, a right to life. But this is enormously question-begging insofar as the conclusion he draws about the morality

McMahan's TRIA, I will contend, is able to defeat the FLOA. After replying to two likely objections to my suggested deployment of this account,
I will conclude that, by persuasively defeating the strongest nonreligious
argument against abortion, this approach largely clinches the case for
the permissibility of early abortion – which is to say, the vast majority of
abortions.[69]

The Future-Like-Ours Argument

In ordinary circumstances, it is uncontroversially wrong to kill paradigm
persons like you or me. Why? According to the FLOA, the fundamental reason – there may be others – is that killing us would deprive us of
valuable futures, which would include all of our personal projects, enjoy-
ments, meaningful activities, and other experiences plausibly believed
to make human life valuable (in ordinary circumstances). This account
of the wrongness of killing explains why we regard killing persons as
such a terrible crime; it also accommodates our belief that death ordinarily harms the person who dies. Meanwhile, it avoids certain difficulties
of other leading accounts of the wrongness of killing persons, such as
the desire-based account.[70] Further, this approach plausibly implies that,
since killing infants would (normally) deprive them of valuable futures –
futures like ours – infanticide is (at least normally) wrong. By contrast,
accounts that base the wrongness of killing persons on their special moral
status as persons, as beings with the capacity for complex forms of consciousness, notoriously struggle to explain the commonsense conviction
that infanticide is (at least normally) wrong.[71]

Now consider fetuses. A human fetus is an individual that can ordinarily, if permitted to live, develop into a person who has the sorts of
experiences that we value so highly. So, the fetus has a future like ours.
Thus, abortion, like the killing of paradigm persons, is ordinarily wrong.
(Our purposes do not require identifying the circumstances that constitute exceptions to the moral presumption against abortion.) This is a

---

of abortion is unsustainable if the GSA proves sound. So we will consider the latter as a
strategy for undermining the FLOA.

[69] Parts of the discussion that follows draw significantly from the section on abortion (pages
426–35) of my "Identity, Killing, and the Boundaries of Our Existence," *Philosophy and
Public Affairs* 31 (2003): 413–42.

[70] See Marquis, "Why Abortion Is Immoral," pp. 195–6.

[71] A prominent example is Mary Anne Warren, "On the Moral and Legal Status of Abortion," *The Monist* 57 (1973): 43–61.

substantial moral argument against abortion that requires no religious assumptions, avoids equivocation on the moral and descriptive senses of such terms as *human being* or *person*, and avoids other difficulties, as indicated earlier. Again, the FLOA is widely regarded as the strongest moral argument against abortion.

But when does it apply? As argued earlier, each of us comes into existence at the point of unique individuation, about two weeks after conception at the latest. The preindividuation organism is not a being of our kind; rather, it is a precursor to, and therefore numerically distinct from, a human organism of our kind. Thus, terminating pregnancy prior to unique individuation is much like contraception insofar as it prevents one of us from coming into existence rather than killing one of us. The organism destroyed by such very early abortions is, like an unfertilized egg or a sperm cell, not the sort that can have substantial moral status.[72] For, being numerically distinct from any later sentient being or person, strictly speaking it cannot even *become* one of us. Therefore, even if the potential to become a being with substantial moral status itself conferred comparable moral status – at best a highly debatable proposition (which I reject) – the preindividuation organism lacks this potential. Thus, terminating pregnancy at this very early stage is relatively easy to justify. (For the same reason, very importantly, therapeutic cloning and stem-cell research are also relatively easy to justify if embryonic development is discontinued prior to the time of unique individuation.)

Thus the FLOA, plausibly developed, applies to the uniquely individuated human organism but not to its embryonic precursor.[73] The uniquely individuated human organism has a valuable future; if it survives, it will (ordinarily) develop into a being with a life containing the sorts of riches that we plausibly believe to make human life precious. Its precursor has no valuable future. As mentioned earlier, I use the term *fetus* to refer only to the uniquely individuated human organism. The question now is whether the FLOA, applied to fetuses, can be defeated.

---

[72] Such very early abortions often occur naturally. Sometimes, however, intrauterine devices and "emergency contraception" work not by preventing conception but by preventing the zygote from implanting in the uterus.

[73] Intending only to provide a general account showing that nearly all abortions are wrong, rather than a complete ethics of abortion, Marquis does not insist that the FLOA applies from the time of conception: "Some anti-abortionists will allow that certain abortions, such as abortion before implantation . . . , may be morally permissible. This essay will not explore the casuistry of these hard cases" ("Why Abortion Is Immoral," p. 183).

A Leading Reply: The Good Samaritan Argument

Suppose the FLOA established a fetal *right to life* – by which I mean a moral protection of fetal life equal in strength to what paradigm persons are generally believed to enjoy. What would follow from this? It may appear to follow from a fetal right to life that abortion is (perhaps with some exceptions) morally impermissible. One of the most ingenious arguments to be developed in the abortion literature, *the Good Samaritan Argument* (GSA), purports to show that abortion is permissible *even if fetuses have a right to life*. Introduced by Judith Jarvis Thomson,[74] the GSA has been recently developed and refined by David Boonin.[75] I take Boonin's contribution, which begins with Thomson's violinist case, as representative.

In the famous thought experiment, you wake up and find yourself in a hospital, hooked up to a violinist who can survive his kidney ailment only if you, who alone have the right blood type, remain in the hospital bed for nine months with the violinist attached to you. He undoubtedly has a right to life. To unplug him in less than nine months would guarantee his death. It nevertheless seems obvious that you are not obliged to undergo such an extensive burden to save his life. For while being such a good Samaritan would be praiseworthy, going to such lengths to assist another (whom one did not consent to help) is well beyond the call of duty. It's not that your rights to liberty and bodily integrity trump his right to life; rather, there is no conflict of rights because the violinist's right to life *does not encompass a right to use your kidneys*. The GSA claims that, even if the fetus has a right to life throughout gestation, unwanted pregnancy is relevantly similar to the scenario involving you and the violinist: Terminating pregnancy with abortion, like disconnecting the violinist, is permissible.[76]

The most promising strategy for critics of the GSA is to identify a morally relevant disanalogy between pregnancy and the hospital scenario. Boonin argues painstakingly against alleged disanalogies. Rather than review his arguments here, let me just say that he persuades me, for example, that the pregnant woman has not tacitly consented to assume the burden of continued pregnancy even if she voluntarily had sex and elected not to use birth control. He also persuades me that the killing–letting die

---

[74] "A Defense of Abortion," *Philosophy and Public Affairs* 1 (1971): 47–66.
[75] *A Defense of Abortion* (Cambridge: Cambridge University Press, 2002). Boonin also draws heavily from Frances Kamm's work. See especially Kamm, *Creation and Abortion* (New York: Oxford University Press, 1992).
[76] Boonin, *A Defense of Abortion*, pp. 135–9.

distinction, motivated by the contrast between aborting a fetus and disconnecting the violinist, will not by itself support a relevant disanalogy. Nevertheless, I think his defense of the GSA (which I believe improves on Thomson's defense) *might* break down.

According to the *responsibility objection*, in nonrape cases – that is, where a woman has had sex voluntarily and becomes pregnant – she (along with the biological father) is responsible for the situation in which the fetus needs life support. Boonin distinguishes two senses of responsibility: (1) One is responsible for the fact that A exists, and (2) one is responsible for the fact that, given that A exists (anyway), A needs your help. In the violinist case, you are not responsible in either sense; similarly with pregnancy due to rape. In nonrape cases, a woman is responsible in sense (1) – she helped to bring the fetus into existence – but not in sense (2), because the fetus could not have existed without the woman's assistance.

Boonin has us consider cases in which one is responsible for A's existence in sense (1) but in a different way – by extending rather than creating a life – and which vary as to whether sense (2) also applies.[77] Suppose, for example, that you, A's doctor, saved his life seven years ago in the only way possible: by administering a drug that cured his disease but would predictably cause kidney failure seven years later. Now, as you expected, you alone can rescue him from his new ailment by giving him the use of your kidneys for nine months. Presumably, you have no obligation to accept this burden. Here you are responsible for A's present existence, since you saved him, but not for his neediness, given that he exists, since he could not have survived to the present day in a nonneedy state. According to Boonin, "where you are responsible in sense (1) and not in sense (2) for the fact that another now stands in need of your assistance . . . the individual in need has not acquired the right to your assistance."[78] According to Boonin, the responsibility objection therefore does not prevail in nonrape cases and, obviously, it doesn't apply in rape cases. It would follow that the GSA establishes the permissibility of abortion – assuming, again, that the responsibility objection presents the strongest challenge to this argument.

But does this reasoning really defeat the responsibility objection? Stressing the special relationship between parent and child, McMahan states a reason to be skeptical: "[I]t is very hard to believe that it is permissible to kill one's own child in order to avoid the burden of

[77] Ibid., pp. 172–5.
[78] Ibid., p. 174.

providing the aid that one has caused it to need."[79] Once again, we are assuming, both for the sake of argument and because the FLOA remains standing, that the fetus has a right to life, as a child does. And, except for hysterectomy or hysterotomy (which pose special risks to the pregnant woman), current methods of abortion involve killing the fetus. Finally, although the pregnant woman is not responsible for the fetus's neediness in sense (2) – it's false that the fetus would have existed even if she had not caused its neediness – *she and the biological father have through voluntary action caused the fetus to exist in a state of need, even if they conscientiously tried to avoid this result by using birth control.* Boonin's two senses of responsibility present a false dichotomy. Here is a sense (3): One is responsible for the fact that *A* exists in a state in which *A* needs your assistance. In nonrape cases one is responsible in this sense, which *perhaps* grounds a right to one's assistance – or, more minimally, a right not to be killed – at least when *A* (who, we are assuming, has a right to life) is one's own child.

Suppose that you, the doctor in Boonin's case, had seven years ago saved the patient, who now, as predicted, cannot live without the use of your kidneys for nine months. Assume that (somehow) the burden to you is roughly comparable to the burden of a typical pregnancy in terms of causing nausea, impact on mobility, and the like (very substantial but less than confinement to bed for nine months). Finally, assume that the patient is your own child. Whereas in Boonin's version of the case – which suggests no special relationship other than that of doctor to patient – you do not seem obliged to give the patient the use of your kidneys, it is much less clear that you lack such an obligation when the patient is your own child. Moreover, it is, if anything, even more dubious that you may kill your child in order to avoid the burden, comparable to a typical full-term pregnancy, of providing assistance that you have caused her to need.

Here, then, is a reply to the GSA based on the responsibility objection together with an appeal to the special parent–child relationship. If allowing one's child to die in order to avoid the burden one has caused her to need is sometimes permissible when killing would not be, then this reply to the GSA also rests on the killing–letting die distinction – in which case it will not apply to nonlethal methods of abortion. (And, again, any version of the responsibility objection will not apply to rape cases.) Either way, a successful reply to the GSA along these lines would mean that the GSA could justify abortion in only a small range of cases, largely defeating

---

[79] *The Ethics of Killing,* p. 398. His supporting argument responds to the GSA in general, not to Boonin's handling of the responsibility objection.

the ambition of this famous defense of abortion. My cautious conclusion is not that the GSA clearly fails, but that there are substantial grounds for doubting that it succeeds.

But we have assumed in discussing the GSA that the fetus has a right to life. Does it? Or can the FLOA be convincingly undermined?

A Successful Reply: The Appeal to Time-Relative Interests

Note that the FLOA makes two crucial assumptions. First, it assumes that each of us was once a presentient fetus. I have defended this assumption with the biological view of human identity. Second, the FLOA assumes that identity is the sole basis for rational prudential concern, for what matters in survival, suggesting that evaluation of a fetus's future should assume a whole-lifetime perspective. From this longitudinal perspective, abortion (ordinarily) entails an enormous loss, the loss of a valuable future – with no discounting of this loss. On the basis of this prudential claim, the FLOA infers that abortion is morally comparable to killing paradigm persons insofar as both acts deprive individuals of their valuable futures.

But from a whole-lifetime perspective, the younger one is, other things equal, the more of a valuable future one loses from death. Thus, the aforementioned prudential claim implies, counterintuitively, that a presentient fetus is harmed more by death than is an infant, who is harmed more by death than is a ten- or twenty-five-year-old (even if we hold that such prudential differences do not matter to the ethics of killing and assign each individual an equal right to life). But, as McMahan notes, precisely the opposite seems true: Other things equal, the ten- or twenty-five-year-old loses more from death than does an infant, who loses more from death than does a presentient fetus.[80] This suggests that the harm of death is a function not only of lost opportunities for valuable future experiences, but also of the way in which one is psychologically "invested" in, or connected with, one's future. McMahan's TRIA explains such judgments about the harm of death, and it can illuminate the morality of abortion.

The TRIA was introduced and defended in Chapter 5, where the issue was the authority of advance directives in cases of severe dementia. The prospect of significantly illuminating the abortion issue by appealing to

---

[80] Ibid., pp. 270–1.

the TRIA warrants further attention to this account. Our question is how we should understand the harm of death for a particular being.

As explained in Chapter 5, the TRIA discounts the importance of death to its victim, at the time of death, for any weakness in the psychological unity that would have connected the victim at that time with himself in the future. The degree of psychological unity in a life, or over a stretch of time, is a function of the richness, complexity, and coherence of the mental life that is carried forward over time. When the psychological unity that would have bound an individual at the time of death to himself in the future, had he lived, is weak, death matters less prudentially – for that individual – at that time. (In Chapter 5, I argued that this discounting is rationally optional, not rationally necessary, in the case of autonomous individuals thinking prospectively, but this point is irrelevant in considering fetuses' interests.) This suggests that a turtle's death does not prudentially matter very much to him, when he dies, because the turtle's mental life is psychologically not very unified over time.

However, one might wonder why, in assessing the harm of death, we should focus on a being's time-relative interest in remaining alive rather than the individual's interest as understood from a whole-lifetime perspective. The TRIA asserts that when there is a significant divergence between what is best for *A* from a whole-lifetime standpoint and what is in *A*'s time-relative interests, we should favor the latter standard. Why? As explained in Chapter 5, the reason is that the TRIA best explains certain comparative intuitions about the harm of death. It helps to explain why a person seems to lose much more from death than a turtle loses. It also provides the only plausible explanation of why death seems to harm an infant less than a ten- or twenty-five-year-old rather than vice versa. This point merits expansion.

The harm of death for an infant seems intermediate between the harm of death for a person and the loss of value of someone's never coming into existence.[81] A conception that never took place entails a loss of enormous possible future good, that of an ordinary human life, but without a victim, the loss is impersonal and therefore of little or no importance. When a person dies, the amount of good lost may well be less than in the nonconception case – since, having already lived some years, the person loses less than a lifetime – but there is a victim, the person. Equally importantly, the victim would have been psychologically deeply connected to herself in the future had she lived. In the case of an infant's death, a

---

[81] McMahan develops this point very lucidly (ibid., pp. 170–1).

great deal of good is lost – on average, more than a dying person loses – and there is a victim, unlike in the nonconception case. But the infant is psychologically only weakly related to herself in the future. If psychological unity did not discount the loss to the victim, the infant would typically lose a good deal more than the ten- or twenty-five-year-old. Discounting the lost good in accordance with the TRIA explains our judgments here.

Let us apply this account to early abortion, the killing of presentient fetuses. No later than two weeks after conception, there is a human organism with a valuable future. This is what is right about the FLOA. *But the utter lack of psychological unity between the presentient fetus and the later minded being it could become justifies a radical discounting of the harm of the fetus's death.* For the proper basis for assessing the harm of death to the presentient fetus is its time-relative interest in remaining alive. Either of two plausible ways of understanding this time-relative interest justifies a major discounting of the harm of death.

On the view I am inclined to accept, because the fetus is identical to the later personal human animal, it has *some* (time-relative) interest in remaining alive, but its interest is very weak, much weaker than yours or mine, due to the absence of psychological unity. Another possibility within the biological approach is to drop the claim that bare identity (with psychological life only in the future) is a basis for prudential concern – in which case the presentient fetus would have *no* interest in remaining alive. Intuitions are likely to clash about which prudential claim is more reasonable. Either way, even if the presentient fetus has an interest in remaining alive, it would be too weak to ground a right to life, so the interests of the pregnant woman or her family could easily justify abortion. Since the vast majority of abortions involve presentient fetuses, this application of the TRIA, if sound, is enormously important.

Two Challenges to This Approach

I have argued that the appeal to time-relative interests constitutes an adequate reply to the FLOA, the strongest argument opposing abortion. Two basic strategies for undermining this approach would be (1) to argue that the TRIA itself is implausible or insufficiently motivated as a theory or (2) to argue that some of its implications are so implausible that there must be something wrong with the account itself. I have already attempted, in Chapter 5 and in this chapter, to argue that the TRIA is a plausible, well-supported account. Let us turn now to the strongest challenges to the TRIA of which I am aware, both of which concern its alleged implications.

Implications Regarding Wrongful Handicap. Suppose a woman frequently uses crack cocaine and abuses alcohol while pregnant and carries the fetus to term. As a result of this substance abuse, the child who comes into the world suffers from numerous disabilities. Assuming the woman was not forced to use these substances, we judge that her behavior is morally indefensible. But the TRIA apparently suggests that we should evaluate the harm done to the fetus during pregnancy on the basis of his *time-relative* interest in being healthy. Now, since the presentient fetus is psychologically cut off from the child he will become, his time-relative interest in later being healthy is extremely weak, just as his time-relative interest in staying alive is extremely weak. How, then, to explain our judgment that the woman's behavior is seriously objectionable – and in individual-affecting terms since the handicapped boy is a victim of her substance abuse?

We can plausibly explain this judgment within our theoretical framework. Because the woman decided against abortion, her fetus has not only a present time-relative interest in being healthy but also many future time-relative interests in being healthy – including when, as a person, he is deeply psychologically unified over time. Because all of these time-relative interests count, the overall harm of the avoidable handicaps is very great, and certainly sufficient to support our moral criticism of the woman's behavior. In examining abortion, by contrast, we count only the fetus's present time-relative interest to live, because an aborted fetus will never have time-relative interests while psychologically deeply unified. Thus my deployment of the TRIA in defending early abortion is consistent with our judgment of wrongful handicap.[82]

Implications Regarding Infanticide. A second, more formidable challenge concerns possible implications for infanticide. According to the TRIA, an infant ordinarily loses more from death than does a presentient fetus – but also loses less than a paradigm person does. An infant has a psychological life, whose unity deepens with time, but this unity is less than that characterizing a paradigm person's psychological life. There is reason to worry, therefore, that the TRIA does not support the right to life that we are inclined to attribute to infants. In this discussion I

---

[82] Although he provides roughly the same argument in the case of late abortions as I just provided in the case of early abortions (ibid., pp. 280–3), McMahan does not apply this reasoning to early abortions, thinking that his embodied mind account demonstrates that the presentient fetus is not "one of us."

assume, with the present challenge, that paradigm persons have a right to life that may be overridden only in certain carefully delineated circumstances (e.g., self-defense, just war, perhaps voluntary active euthanasia [including, if the traditional standard of death is correct, that involved in harvesting vital organs – see Chapter 4]). Does my approach imply that infants, who are not paradigm persons, lack such a right to life?

Those who are relatively liberal about active euthanasia are likely to accept infanticide in those rare cases where death would be better than continued life *for the infant herself* – that is, when death is in the infant's best interests (as in wrongful-life cases). But the TRIA may imply the much more radical view that infanticide is justified fairly often even when the infant has a time-relative interest in continuing to live. Perhaps discounting this interest in accordance with the infant's diminished psychological unity would permit the former to be overridden by the interests of family members or the broader society in a rather wide range of cases.

But my defense of early abortion does not support radical openness to infanticide. Presentient fetuses and infants are differently situated in several ways that justify a very strong presumption against infanticide but very little presumption against early abortions. Let me explain.

First, *infants are sentient, whereas presentient fetuses, by definition, are not.*[83] Only sentient beings have experiential interests such as interests in avoiding pain, distress, and any other aversive experiences they are capable of having. The process of being killed can be unpleasant for an infant but not for a presentient fetus. Then again, because infanticide can be performed in ways that are virtually painless for the infant, this consequence of sentience has little importance in the present context.

Another consequence of sentience is far more significant. Earlier I noted my inclination to accept the controversial thesis that even a presentient fetus has a time-relative interest, though a very weak one, in continuing to live. If that is right, then while the infant's having a psychological life per se might not entail a stronger time-relative interest in continuing to live than the presentient fetus has, any unification of this psychological life – that is, any carrying forward of this psychological life *over time* – will strengthen the time-relative interest. (Again, the degree of psychological unity in a life, over some stretch of time, is a function

---

[83] More precisely, nearly all infants are sentient. Anencephalic infants, lacking a functioning cerebrum, are not and never will be sentient. But infanticide in their case – for example, in an effort to remove vital organs in order to help other infants – may well be justified insofar as beings who lack even the potential for sentience or consciousness lack interests and therefore cannot be harmed in any morally significant way.

of the richness, complexity, and coherence of the mental life that's carried forward over time.) Whereas psychological life may develop very little between the time at which a fetus becomes sentient and the time of birth (perhaps just a greater capacity to feel sensations due to neural growth), postnatal psychological life develops very quickly due to the infant's exposure to the bustling, varied, and highly social world outside the womb. The infant's psychological life rapidly becomes more complex and unified over time, entailing a stronger time-relative interest in continuing to live. On the approach that appeals to the TRIA in justifying early abortion, this difference between presentient fetuses and infants is significant.

This difference is even greater if I am wrong that presentient fetuses have a time-relative interest in continuing to live. If they do not, then sentience per se will make a more significant difference because the emergence of psychological states and capacities will mark the emergence of a time-relative interest in continuing to live. In that case, whereas the presentient fetus has *no* time-relative interest in continuing to live, an infant clearly has some such time-relative interest, providing a reason for treating infanticide as more morally problematic than early abortion.

Second, *birth signals the infant's entry into a social world in which the presentient fetus cannot participate.* It is true that a pregnant woman and other people can talk to her fetus, see it on a sonogram, feel its movements, and introduce it to Bach's cello suites. But these "social interactions" bear no honest comparison to the social interactions in which an infant participates, even on her first day of postnatal life. The new baby is held, dressed, fed, spoken to frequently, changed, cleaned, and engaged in eye contact. Anyone present can behold her. Soon after birth she participates actively in social interactions with facial expressions, vocalizing, and reaching. There is a profound sense in which an infant is socially present in a way that no fetus is. And the infant's social presence helps to account for the fact that both our legal institutions and common morality regard newborns as having full moral status, including a right to life, while regarding fetuses' moral status with great uncertainty.

One might reply that only persons – beings with the capacity for sufficiently complex forms of consciousness – are rightly regarded as having full moral status, suggesting the inadequacy of both our legal institutions and common morality in how they regard newborn infants. Accordingly, one might propose that we revise both the law and our moral perceptions to reflect the fact that prepersonal infants lack the moral status of persons.

But this proposal is problematic for several reasons. First, the only un-controversial thesis about the moral status of persons is that personhood is *sufficient* for full moral status. As discussed in Chapter 1, the thesis that personhood is *necessary* for full moral status is increasingly contested and, more importantly, is genuinely contestable.

Second, we have no confident way of saying when the developing hu-man being becomes a person. Indeed, because the concept of person-hood is significantly vague (see Chapter 1), there is no *line* dividing per-sons and nonpersons; there is instead a gray area separating the paradigm cases on each side. To be sure, newborns are not persons in the relevant sense, whereas ordinary speaking two-year-olds clearly are. But in the continuum from early postnatal life to paradigm personhood, there is a range occupied by borderline persons.

One might reply that we could conservatively draw a line that would protect both persons and borderline persons, counting, say, only new-borns one month old or less as lacking the rights attributed to persons. While possible in principle – assuming we grant the contestable thesis that personhood is necessary for full moral status – this sort of policy seems im-possible in practice, providing the third argument against it. The reason such a policy seems impossible in practice has everything to do with the newborn's participation in the social world of persons. For it is simply a fact that the sensibilities of a vast proportion of the public, surely an over-whelming majority, would be greatly offended by the proposed openness to the killing of newborns. Presentient fetuses are in a different position. While an impressive minority of persons are opposed to abortion in all or almost all circumstances, nearly everyone, it seems, opposes infanticide – the killing of human beings who are meaningful participants in our social world – in all or almost all circumstances. This constitutes a pragmatic reason for not revolutionizing our laws and common moral judgments about infanticide. We have, then, another reason to doubt that the TRIA has unpalatable implications regarding infanticide.

A final salient difference between infants and presentient fetuses is this: *The personal costs to the pregnant woman of continuing an unwanted pregnancy are typically, and almost inherently, much more substantial than the personal costs to the biological mother or any other individual of an infant's continuing to live.*[84] A pregnancy involves the fetus's continued, increasingly burdensome use of a woman's body (a point underscored in the GSA). Pregnancy, child-birth itself, and recovery from childbirth entail a great deal of pain and

---

[84] McMahan develops this point persuasively (*The Ethics of Killing*, pp. 344–5).

discomfort, can have significant psychological costs – especially if the pregnancy was unwanted – and may substantially interfere with the pregnant woman's lifestyle and opportunities. By contrast, *having borne* a child does not necessarily involve continuing use of the mother's body (though breast-feeding, if chosen, is a lesser form of body use). Moreover, the possibility of giving up a child for adoption means that the personal costs of a continuing infant life for the biological mother may be relatively small and transient. These points relate to the simple fact that pregnancy involves profound use of a woman's body in a way that having given birth does not. Further, even if a pregnant woman has an enlightened, considerate partner, a supportive broader family network, or other forms of support, there are drastic limits to what a partner, family, or society can do to lessen the burdens of pregnancy for the biological mother – due, again, to what pregnancy inherently involves: continued profound use of a woman's body. Once an infant has been born, even in the vast majority of cases where she is not given up for adoption, the possibilities for distributing the responsibilities and costs involved in child rearing are extensive. While the differing burdens of pregnancy and of having borne a child do not seem directly relevant to the moral status of the fetus and infant, they bear significantly on a crucial moral factor: the costs to the prospective biological mother of granting fetuses or infants a right to life.

For all of these reasons, then, I conclude that appealing to the TRIA in justifying early abortion does not have unpalatable implications regarding infanticide. Infants, but not presentient fetuses, are properly regarded as having a right to life. Inasmuch as the most widely accepted justifications for (intentional) killing – such as self-defense and just war – do not apply to infants, it is an open question whether they should ever intentionally be killed. If so, that will always or nearly always be in cases where continuing to live is contrary to an infant's best interests. (I have left open the question of whether active euthanasia, which involves killing, and not only forgoing treatment in allowing someone to die, is justified at the level of public policy.)

Conclusion

Having defended the abortion of presentient fetuses, I will not attempt to extend the analysis to sentient fetuses. Sentience, which emerges between five and seven months after conception, is morally significant, as explained earlier. Fetuses become viable roughly six months after conception – that is, at very roughly the time they become sentient. Viable fetuses *could be* infants, yet their continued presence within the mother's body cannot be ignored; whether viability carries any special moral significance is debatable. I find the issue of late abortion more puzzling and difficult to resolve than the issues of early abortion and infanticide. Rather than addressing late abortion here, I will rest content with this section's defense of early abortion – the only sort that most girls and women who consider having an abortion ever consider.

Index

Printed in the United States
By Bookmasters